Seismic Exploration Fundamentals
Second Edition

Seismic techniques for finding oil

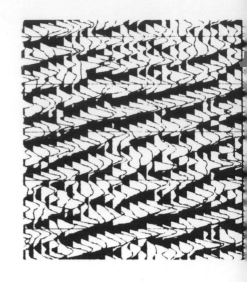

Seismic

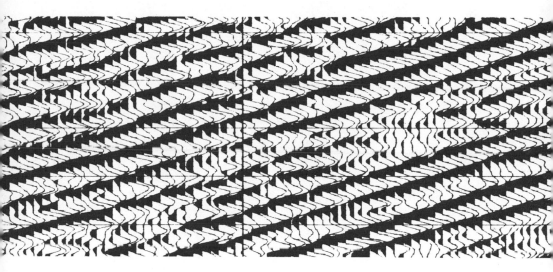

Exploration
Fundamentals

Second Edition

Seismic techniques for finding oil

J. A. COFFEEN

PennWell Books
PennWell Publishing Company
Tulsa, Oklahoma

Copyright © 1986 by
PennWell Publishing Company
1421 South Sheridan Road/P.O. Box 1260
Tulsa, Oklahoma 74101

Library of Congress Cataloging-in-Publication Data

Coffeen, J. A.
     Seismic exploration fundamentals.

     Bibliography: p.
     Includes index.
     1. Seismic prospecting.   2. Petroleum.   I. Title.
TN271.P4C63   1986   622'.18282   85–19112
ISBN 0–87814–295–9

Printed in the United States of America

1 2 3 4 5 90 89 88 87 86

Contents

9

Interpretation Three Ways *252*

10

Other Seismic Methods *282*

11

Old Data and New *295*

12

Preface

This is a second edition. The first came out in 1978. Geophysics has changed so in the intervening years, that a new version of the book is necessary. In particular, 3-D exploration, VSPs, shear wave exploration, and interactive interpretation were, at most, of minor importance in oil exploration at the time of the first edition. In this new edition these topics are stressed. Also, other changes have been made throughout the book.

This is a book on using seismic techniques to find oil. It is not on mathematical and theoretical aspects of seismic exploration.

It is a manual designed for:

Managers, who need an overall knowledge of seismic exploration, so they can work effectively with geophysics and geophysicists;

Geologists, whose work is closely related to geophysics;

People in specialized fields of geophysics—data processors, field personnel, etc., who would like an understanding of how their specialties fit into the exploration effort;

Students, who want to learn what seismic exploration is like in actual practice;

Finally, new employees with a theoretical background in geophysics, who need to know how that background is used in finding oil.

The theoretical side of geophysics is vital. Seismic exploration began because theoretical principles, particularly Huygens' Principle, were developed. It is now progressing at an almost headlong rate, partly because of research being carried out by oil companies, seismic contractors, and universities.

This book is concerned with an entirely different problem—that of finding oil in specific situations. This calls for the exploration attitude, competitive, affected by deadlines, including many aspects of the situation simultaneously.

Acknowledgments

Thanks are due to a number of people and organizations—the companies and individuals that contributed illustrations and examples, the people who answered my long lists of questions about how things have changed since the first edition, and the friends who read parts of the text and commented on them.

CHAPTER 1

What It's All About

Conventional seismic exploration for oil is searching underground for oil by use of sound. The word "seismic" refers to vibrations of the earth. It includes both earthquakes and sound waves that penetrate well through the earth. This sound is in the frequency range of about 10-100 cycles per second. People can hear sounds from about 20 to 20,000 cycles per second, with some individual variation. So the seismic frequency range is right around the deepest tones people can hear. The sound used in seismic exploration is mostly so weak, though, that it isn't audible without amplification.

A sound in the seismic range of frequencies will travel deep into the earth and return, bringing back information with it. The depths investigated are from about 1000 feet to as much as 10 miles.

Two types of seismic exploration, reflection and refraction, are based on two ways that the sound is returned to the surface. Sound can be reflected (echoed) from a rock layer, or it can be refracted (bent) along a layer and then up to the surface. Refraction was the first type of seismic exploration to be commonly used, but reflection has now supplanted it in all but a few specialized uses. So reflection will be the method described in most of this book. Reflection exploration is conducted by artificially producing a sound at or near the surface of the earth, recording the echoes from underground, and analyzing them.

Of course, it isn't as simple as that. The sound has to be produced in just the right way, the recording has to be done carefully and correctly, there are many things to be monitored and kept track of.

In conventional seismic exploration and its major variations, here is what happens.

Lines are drawn on a map to indicate where seismic investigations are to be conducted. On the ground, the locations and elevations of those lines are surveyed, and points along the lines are marked with stakes or tags or something, to indicate where sounds are to be made and received.

Small sound detectors are placed on the ground, along the lines of marked positions. The detectors are connected by electrical cable, optical fiber cable, or radio to tape recording equipment. Sounds are made by vibrating or dropping a weight, by firing an explosive, or by other means. The sound reflected from underground is recorded on magnetic tape. This is repeated along the line, and along the other planned lines (Fig. 1-1).

At sea, the investigation is conducted from a boat. The surveying is done by radio. The detectors are trailed behind the boat, and the sound is usually generated by the release underwater of compressed air. The recording is done in the same way as on land.

The tapes are sent to a processing center, where computers rearrange the information, modify and enhance it on a succession of other tapes to make it provide more useful information about underground layers of rock. The data on the final tapes is displayed in visual form on film, paper, or video screen. The data from one line is displayed as a cross section of the upper several kilometers of the earth, or the data from one moment of time for an area is displayed as a horizontal section through the earth.

The displays are used to determine the shape and other characteristics of the layers of rock underground. This is done by marking on paper displays or by giving instructions to a computer. The configuration of a rock layer is plotted on a map and contoured. Possible oil and gas traps may be indicated by the contours and by other information obtained from the reflections. Wells may be drilled into those possible traps, and oil or gas may be found.

Terms

People engaged in a specialized activity tend to develop expressions that make it easier for them to discuss the distinctive characteristics of that activity. Helpful as those expressions are to the specialists, though, they make it more difficult for outsiders to understand. Sports, arts, sciences, farming, geophysics—each has its own terminology that makes it easier to talk about the topic if you know the terms, but difficult to understand if you don't. So, throughout this book, some stress will be put on the specialized terms used in geophysics. A few will be introduced now. These first terms refer to the description just given of a seismic operation.

The arrangement of locations for a seismic investigation, the lines drawn on a map, is a seismic program. The act of deciding where to draw the lines is program planning.

The people who do the work in the field are a seismic crew, or

Credit: J.A. Thomas

Fig. 1-1

seismograph crew, also a doodlebug crew, and the individuals doodle-buggers.

Locating the positions on the ground and determining their elevations is surveying, not really a specialized geophysical word.

Producing the sounds is called shooting, from the earlier time when explosives were the only way the sounds were produced. Also the whole operation in the field is called shooting.

The shooting of an area is sometimes called a seismic survey. This is a little confusing, because it sounds like it might refer only to the surveying of the locations.

The sound itself is referred to as seismic energy, seismic waves, or just energy. The more common sources of energy are explosives in a hole; vibrator, which initiates sound by vibrating a weight pressed against the earth; weight drop, dropping a heavy weight on the ground; air gun, which releases compressed air in water.

A hole in which an explosive is fired is a shot hole. Its geographical location is a shot point. By analogy and habit, the location at which sound is generated by any means is often called a shot point. And the sounds are produced at such close spacing that sometimes "shot point" means, not every point shot, but only those used in mapping. For example, when compressed air is used, each point is called a pop, and every tenth, or twentieth, or some number, pop is arbitrarily chosen to be mapped and called a shot point.

The line of continuous shot points is a seismic line, or just a line. Shooting arranged in widely-spaced lines is 2-D, two dimensional, shooting. Shooting arranged to give closely spaced information in every direction is 3-D, or three dimensional, shooting.

The sound detectors are called geophones, seismometers, detectors, or in field terminology, jugs, seises, pickups, phones. The ones designed to be used in water are usually called hydrophones. From communication theory, any means of producing seismic energy is called a source, and a geophone or group of geophones, a receiver.

The electrical cable connecting the geophones to the recording instruments on land is just called a cable. The cable used at sea is usually spoken of as a streamer. Streamers have the hydrophones built in.

The magnetic sound-recording tape is just called magnetic tape or tape. One on which the sound is originally recorded is a field tape. Then data processing produces intermediate tapes and the final tape, from which the visual display is made.

A display of the data as a cross section of the earth is a seismic section, a record section, a vertical section, or just a section. A horizontal display is referred to as a time slice, or horizontal section.

A vertical section is made up of alignments of data called traces. A trace may be a wiggly line, a lineup of light and dark patches, or some other form.

The echoes from continuous layers underground are reflections. The alignment across a section that represents a reflection is also called a reflection. A horizon is the surface the reflection bounces from, and also the reflection itself, particularly one that extends over a fairly large area. The time in which sound goes down to a horizon and back up is the reflection time.

The obtaining of information about rock layers from the seismic displays is interpretation. It is done by marking reflections on the sections and slices, and then mapping that information. When it is performed on a computer it is interactive interpretation.

These terms are just a few to get started. Others will be taken up with specific topics.

Echo Ranging

Seismic exploration is an echo-ranging system. That is, it includes both transmitter and receiver, and determines distances to objects by measuring elapsed time between sending energy out and receiving it bounced back from the object. Note that earthquake seismograph is not echo ranging, as it does not involve transmitting energy, but only receives energy from other sources.

There are several echo-ranging systems that are well known.

The best known is radar, in which radio waves are emitted, and then detected when they are reflected back from something. The measurements can be taken at many angles and the results can produce a map on a video screen. The radio waves travel at the speed of light, so the time of travel is extremely short, and must be measured very precisely. Radar is used in weather forecasting to determine the shapes and locations of clouds, and in navigation of aircraft and ships to produce a map of shorelines, other traffic, etc.

Bats have an echo-ranging system that uses sound waves. Two statements are often made about their system—that they have a built-in radar, and that they had it a million years before we did. Well, it isn't exactly radar, as it uses sound rather than radio waves, but it is used for navigation in much the way radar is. The bats make many high-frequency chirps and determine, from the reflected chirps they hear, the locations of obstacles and of the insects they want to catch.

A water depth recorder also emits sound and detects the reflected sounds. It uses the information to make a strip chart of the water bottom along a vessel's course.

Reflection of Sound

Sound is a longitudinal vibration of matter. That is, it is a series of compressions and decompressions expanding outward in all directions from its origin. Its velocity depends on the material that is being expanded and contracted. The state of the material (solid, liquid, or gas), the type of material, and to a lesser extent its pressure and temperature, control the velocity of expansions and contractions through it, the velocity of sound in it. When this compressional wave meets a different substance, it sets the second material vibrating too, but at the different velocity of that second substance. The vibration moves forward through the second material, but also its vibrating causes the first material to vibrate, sending expansions and contractions back through it. The vibration going back is an echo, or reflection, of the sound.

The reflections occur at any change in the velocity of the vibrations, but are stronger where the change in velocity is greater. The velocities of sound that seismic exploration is concerned with are about 1000 ft/sec in air, about 4900 ft/sec in sea water, and from around 1500 to 26,000 ft/sec in the earth. The 1500 ft/sec is for loose soil at the surface, and the 26,000 ft/sec applies to some deeply buried hard rocks. Harder rock tends to have faster velocity of sound than softer rock. Deeper burial in the earth compresses rock and makes it harder.

Sedimentary basins (where oil is most likely to be found) are composed of layers of different kinds of rock mostly deposited smoothly on one another. A change from one kind of rock to another is a change from one velocity of sound to another, a velocity interface. There is a reflection of sound, a seismic reflection, at each of these interfaces, with more of the sound reflected at the greater changes. Density of the rock also affects the amount of reflection, but velocity has the greater effect (and velocity is largely a result of density), so geophysicists often speak of velocity interfaces without mentioning the density differences.

The expanding compressional waves are more or less spherical, so the behavior of seismic energy can be sketched in cross section by drawing arcs with a compass (Fig. 1-2). However, it is also useful to consider ray paths, that is, lines representing the route of the bit of sound that goes from source down to point of reflection, and up to geophone (Fig. 1-3). A ray path is the route of one part of the wave front (Fig. 1-4).

A basic geometrical fact in seismic exploration is that a thing rebounding from a surface leaves that surface exactly as steeply as it approached, or "the angle of incidence is equal to the angle of reflection" (Fig. 1-5). This applies to a tennis ball bouncing from a court, a light reflecting from a mirror, sound reflecting from a rock layer.

The tennis ball follows a single path, but sound goes outward in all directions. The ray path of sound is comparable to the route of the tennis ball. In a simple situation, with the ground and the reflecting surface both flat and uniform in makeup, and the receiver some distance from the source, the ray path will be a straight line from source to reflecting surface and up to receiver—two straight lines meeting at

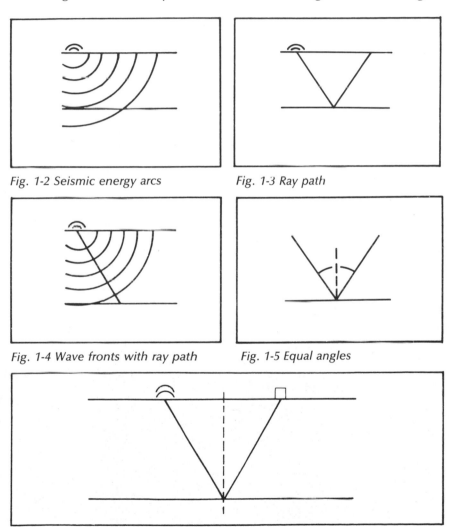

Fig. 1-2 Seismic energy arcs   Fig. 1-3 Ray path

Fig. 1-4 Wave fronts with ray path   Fig. 1-5 Equal angles

Fig. 1-6 Path of reflection

equal angles to the perpendicular (Fig. 1-6). So the sound is reflected from a point on the rock layer directly beneath the midpoint between source and receiver.

This is a basic principle of seismic exploration. *A seismic shot yields data from midway between source and receiver.*

The rule just stated is true, though, only when both the ground and the reflecting layer are flat. They usually aren't. So what does happen? Suppose the reflecting horizon dips.

Let's first consider the situation when the source and receiver are at about the same place, so the sound returns to its point of origin. The angle of incidence and the angle of reflection must be equal. The only way this can be true in this situation is for the sound to strike the layer perpendicularly. This agrees with things in ordinary living too. To throw a ball and have it return to your hand, you must throw it so it strikes something at a right angle. To have it bounce back from the floor, you throw it straight down (Fig. 1-7). But if you throw at a leaning sheet of plywood, then you have to throw outward so the ball hits the plywood perpendicularly (Fig. 1-8).

The same common-sense reasoning applies when the source and receiver are some distance apart. If you want to throw a ball so another person can easily catch it after the first bounce, you throw it to hit the floor midway between you and the other person. And, although this is not as common an experience, if the floor between is tilted, you have to throw at a different angle to compensate.

So, when the reflecting surface slopes, the point of reflection is not beneath the midpoint between source and receiver, but offset some-what. Which way? In the updip direction, up the slope of the tilted surface, as in Fig. 1-9 if the source and receiver are at the same place, or as in Fig. 1-10 if they are some distance apart. In both cases, the angles of incidence and reflection are equal, the way any reflection or bouncing must work.

Fig. 1-7 Level surface

Fig. 1-8 Tilted surface

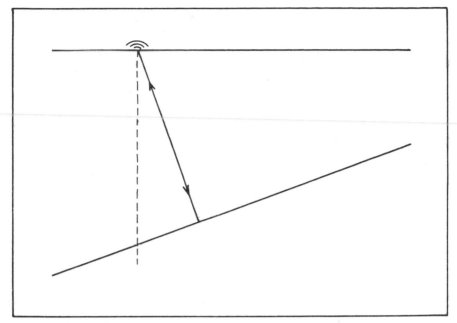

Fig. 1-9 Offset with dip

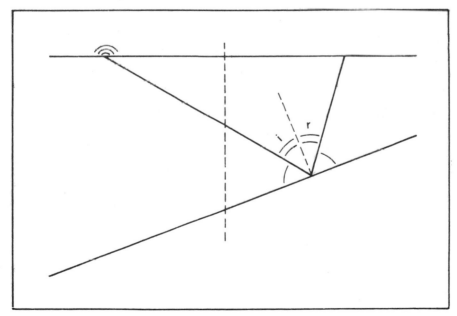

Fig. 1-10 Offset for distant geophone

So reflection usually takes place, not at a point exactly midway between source and receiver, but at a point farther updip. Most rock layers do not dip very steeply, though, so the reflection occurs not far from the midpoint. Therefore, assuming that reflection takes place at the midpoint is a useful rule of thumb.

The main value of seismic reflection in exploring for oil is to provide information on relative depths of a layer. The reflection from a specific layer is recognized along a seismic section, and the reflection times determined at different points on the line. These times are clues to relative depths to the reflector. A time less than normal indicates a point where the reflector is structurally high.

Group

In describing what happens when a source and receiver are in certain positions, they were treated as single points. However, useful seismic data, which is reflected from deep in the earth, is received by a geophone along with considerable "noise", that is, unwanted energy. So it is worthwhile to arrange things so the wanted data is favored, while discriminating against the noise.

Sound that travels direct from the source to the geophone, is almost horizontal for a distant geophone (Fig. 1-11). But the energy reflected from below is nearer to vertical. To favor the vertical energy and diminish the effect of the horizontal, the one geophone that would produce a trace is replaced by a group of geophones, spread out over the ground (Fig. 1-12). The vertical energy strikes all of them at about the

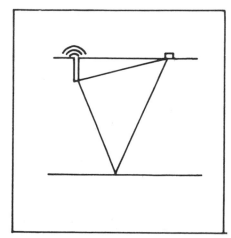

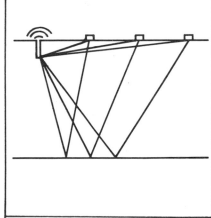

Fig. 1-11 Roughly vertical and horizontal

Fig. 1-12 Geophone group

same time, but horizontally traveling sound strikes the geophone nearest the source, then progresses to the others. The geophones in the group are connected electrically by wires, so they feed data into the cable as though they were only one geophone. The horizontal energy, reaching the different geophones of the group at different times, is transmitted through the cable out of phase. That is, in the extreme case a compression may reach one geophone at the same time a decompression reaches another. When those blend, they tend to cancel each other. So the total horizontal energy transmitted is reduced. Vertical energy, arriving almost simultaneously, is transmitted more or less in phase, and so tends to be reinforced.

Most seismic data is recorded from groups of geophones, rather than single phones. The geophones of a group are spread along the cable for a distance of several feet to several hundred feet.

Special arrangements of geophones in groups are sometimes designed to permit better noise cancellation for a specific situation. Groups are often "tapered", that is, the geophones in the group are unevenly spaced, closer together near the center of the group and farther apart near the ends. (Fig. 1-13). A taper can be designed very specifically for reduction of the noise encountered in a particular area.

There are also geophone patterns that are not along the seismic line, but cover an area. A number of types have been designed—star, several parallel lines, etc. The advantage is that they discriminate against noise from the sides as well as directly from the shot.

The noise-cancelling effect is exactly the same if, instead of several geophones, there are several sound sources. The sources can be used simultaneously, or one source used several times and the data combined later on tape.

If there are several of each—source and receiver—the cancellation of horizontal noise is even greater.

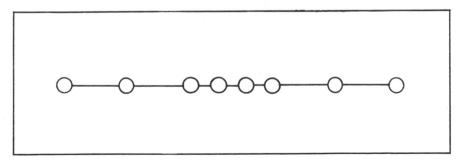

Fig. 1-13 Tapered group

Spread

The geophone groups are placed in a "spread", spaced along an electrical or optical fiber cable, or as telemetry units spaced in line. There may be 96, 120, 240, or some other number of groups for a shot. Each group puts one sequence of electrical signals, corresponding to vibrations of the ground, into the cable. The signals go to the recording instruments, where the information in each group is recorded separately.

The spreads may be from one and a half miles to two miles long, with the groups from 110 to 330 feet apart. Sometimes the groups are longer than the distances between their centers, so adjacent groups overlap. Although the overlap seems paradoxical, the important things are the placing of centers of groups and the noise-cancellation in each group.

The single-ended spread, or single-ender, has the cable laid out in one direction from the shot point. Each geophone or group receives data from a point half way between it and the shot point (Fig. 1-14), so the whole spread gives information from the shot point half way to the farthest geophone.

If a spread is laid out in two directions from the shot point, then the picture is just the same, but with an additional spread going the other way. So subsurface data extends from the shot point half way to each end (Fig. 1-15). This is a split spread.

A line of shot points using either split or single-ended spreads is a continuous line. It can be extended for any distance, and with complete overlap of spreads on the surface, the half-distances of subsurface recorded can be arranged to give complete subsurface information along the line. This is 100% subsurface coverage, or 100% shooting.

An increase from 100% to higher percentages is achieved by the common depth point process, which covers the same subsurface more times, to produce better data.

Common Depth Point

At one time, before magnetic tape recording had come to seismic exploration, there was only one data combining method available. This was called mixing, or, to make it sound more sophisticated, compositing. In its simplest form energy from two adjacent geophone groups was combined into one trace, just by hooking them to the same wire. Thus, the trace was an electrical composite of the data received at the

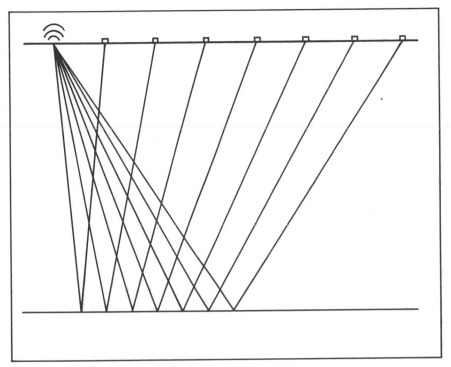

Fig. 1 14 Single-ended spread

two locations. This mixing could take place as described, out where the instruments were connected to a cable leading to the recording truck, or the blending could be done with the same effect in the truck, by connecting the wires that had brought the two separate sets of data to the truck. The first was called a ground mix and the second an instrument mix, or just a mix.

It worked fine for flat data, but tended to discriminate against dip as well as noise. It did help in finding large, gently dipping features.

Then the advent of magnetic tape recording made it possible to combine the data from different shots into one seismic trace, like the way a singer's voice can be superimposed on a tape more than once, to produce a one-person duet, trio, etc.

A method of getting the advantage of combining data without the bad effects of mixing was devised. It is the common depth point method of shooting. It is also called CDP, common reflection point or CRP, common midpoint or CMP. The term "common midpoint" is

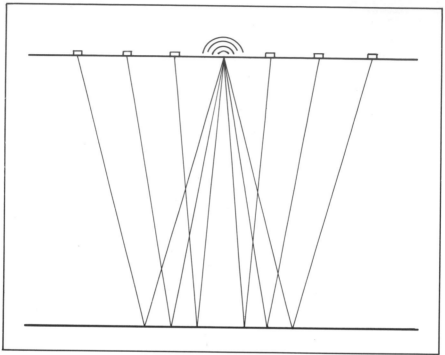

Fig. 1-15 Split spread

technically more correct than the other terms, but for the present CDP is ingrained in the geophysical language, so it will be used in this book. The combining is called stacking. CDP data comes in percentages of stack—1200%, 4800%, etc; or in fold—12 fold, 48 fold, etc.

The principle of CDP shooting is simple. A source and a receiver produce subsurface data from approximately under the midpoint between the two (Fig. 1-16). Another source is placed farther from the center, and a receiver is placed equally far in the opposite direction (Fig. 1-17). Energy from the second source that arrives at the second receiver is reflected from the same point in the subsurface. So sound traveling by two paths gives information from the same point. Part of the longer path can be subtracted to make it the same length as the other, and the two traces stacked, combined into one. The data from the reflection is added, but noise that occurs at different times on the two traces is not added. So the stacking improves the seismic data.

Only 200%, or two-fold, CDP was described just now, but it's easy to see how more routes of energy can be added to increase the multiplicity of stack (Fig. 1-18).

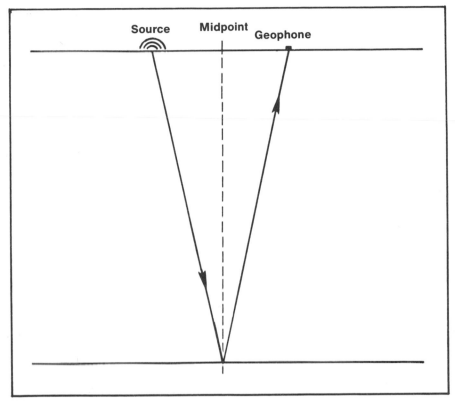

Fig. 1-16 Single source and receiver

In actual practice, CDP shooting isn't done by choosing a spot, and taking a succession of farther shots received by farther instruments. The operation is made more wholesale. A long cable is laid out, with many groups of geophones along it. Then shots are fired at fairly close intervals along that line (Fig. 1-19). The various energy paths are sorted out later in data processing into collections that have the same midpoint.

Offshore, a cable one and a half to over two miles long is trailed behind a boat, and sounds are produced near the boat. Again, the long and short energy paths through the same midpoints are sorted out and put together in processing.

CDP includes some basic assumptions that dictate shooting procedure. For one thing, the geometry applies to points in a straight line. However, a slight deviation from a straight line will yield data that can be successfully stacked. Fifteen degrees is the generally accepted limit to the amount of bend in a line. Where a line cannot be straight at all,

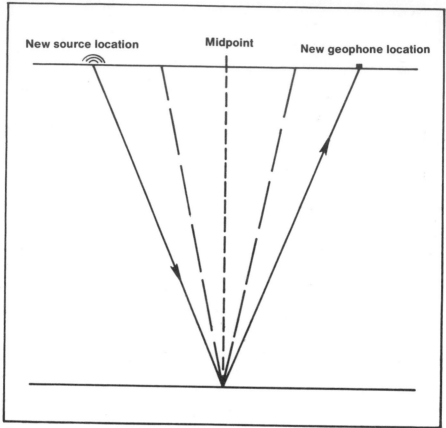

Fig. 1-17 Two sources and receivers

the midpoints can be plotted on a map, and those near enough to each other can be sorted out and combined.

Another assumption is that the reflecting horizons have no dip. Reflection takes place directly below the midpoint between source and receiver only when the reflecting surface is flat. The shift away from the midpoint is not great for small amounts of dip, though, so CDP stacking works well even for most dipping beds. Dip not only shifts the reflection point out from under the midpoint, but shifts differently for different energy paths (Fig. 1-20). If source and receiver are close together, the shift is small. For greater source-receiver separation, the shift is greater. So the depth points are not common. However, the depth points are usually close enough for the "smear" caused by the separation to do less harm than the stacking does good. That is, the stacked data, even with smear, is still better than 100% data.

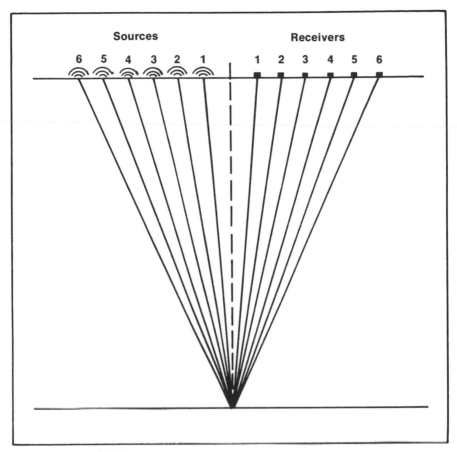

Fig. 1-18 600% CDP

One of the main uses of CDP is in suppressing multiples, that is, reflections that are repeated. Briefly, a multiple is a reflection that bounced off a layer in the subsurface, moved upward, bounced downward from some interface, and then was reflected upward again. The use of CDP to diminish multiples is discussed in Chapter 5, Data Processing.

Line Spacing and 3-D

In shooting an area, the program may consist of only a few lines or it may consist of many, depending on the purposes of the program. If a uniform grid of lines is used, the most important lines are in one direction, generally up and down the dip of the formations. The lines in the

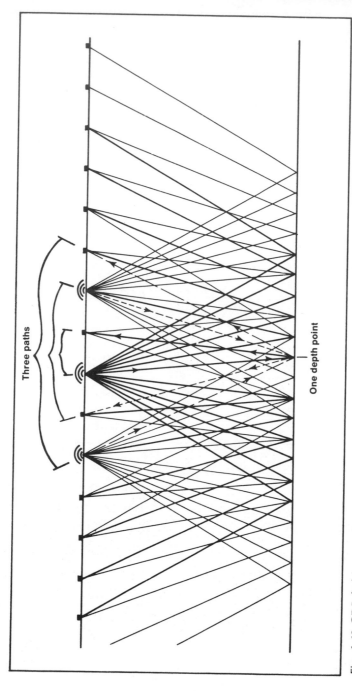

Fig. 1-19 CDP field procedure

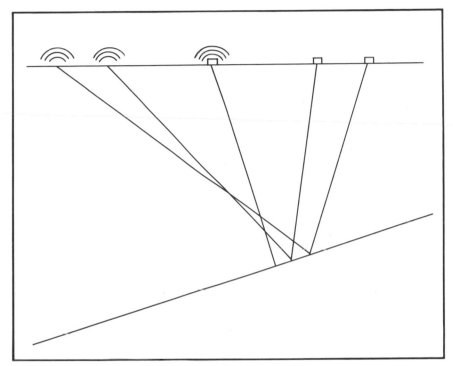

Fig. 1-20 Depth point not common

other direction, along strike, perpendicular to dip, are usually farther apart. The strike lines connect the dip lines, so an interpreter can pick a reflection from one line to another over the area. The strike lines form loops with the dip lines, so the interpretation can be checked when it gets back around to its starting point. Shooting where the prospective features are small in area or shooting over a discovered field, the lines—particularly the dip lines—will be put closer together than normal to detect finer detail.

Each line that is shot records data that will be put together in data processing into a seismic section—a two-dimensional display of the subsurface, like a cross section of part of the earth. This is conventional two dimensional shooting—2-D.

Putting lines much closer together creates a different type of shooting that gathers more complete data. The space between lines may not be much different from the space between geophone groups on a line. Reflections on any section appear continuous (even though there are gaps between traces) because the rock layers don't change much in the distance between traces. So if there is a similarly short distance between lines, the information is similarly continuous from line to line. The data acquired from below the lines forms a three dimensional volume of data, below points that blanket an area. Traces from different lines can be put together to make up a section in any direction, as though a line had been shot in that direction. Shooting in this way is a three dimensional survey—3-D. With the lines so close together, there is no need for intersecting lines to tie them together. So 3-D shooting can be composed of just parallel lines.

Monitor Record

A monitor record is a display of traces from a single shot (Fig. 1-21). Monitor records are played out in the field to check on quality of data being recorded and on how well the recording is being done. The record may be 6 to 10 inches wide and a few feet long. Some monitor records are printed by xerography, the process used in office copiers, and some by other processes. There are traces in the form of wiggly lines, like irregular sine waves, running the length of the record. Each trace was recorded from a different geophone group. The traces are close together, so they overlap onto one another.

There are timing lines crossing the record, to mark time elapsed since the shot, usually one line every 10 milliseconds.

Progressing down the record, from early to later times in the recording, the traces are at first quiet, that is, nearly straight, with only small wiggles, caused by wind blowing across geophones or other slight disturbances. A trace that is quiet for its entire length, maybe because something was disconnected somewhere in recording, is called a dead trace.

The first event, the first noticeable thing that happens at a specific time on the record, is the time break. This is an abrupt burst of energy on one trace, appearing as a sharp wiggle that extends far from the quiet trace position. It is made by the firing of the shot or initiation of other type of source, so it marks the instant the energy is initiated. It is used as a starting point for measuring times on the record. Time to another event is obtained by counting timing lines after the time break, and interpolating points between timing lines. The time break is the starting point for all geophone groups, so it does not belong on any

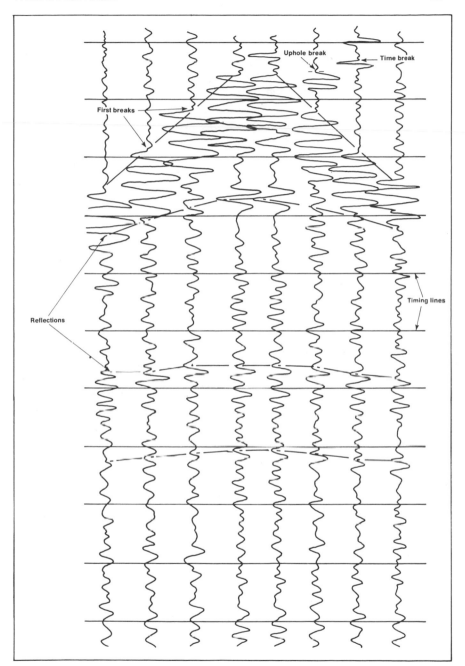

Fig. 1-21 Schematic of monitor record

trace any more than any other. So it is arbitrarily put on some trace where it is not in the way of other information.

On records taken by firing an explosive in a hole, the next event is the uphole break. It is recorded by a geophone on the ground, very near the hole, maybe ten feet away (Fig. 1-22). The first sound to reach this geophone traveled directly up through the earth from the shot. The time measured between time break and uphole break is called the uphole time. The geophone is the uphole jug, shot point seismometer, or SPS. On records taken on land with sources used on the surface (no shot hole), and on offshore records, there is no uphole time.

Next in time on the record come the first breaks. They are the first wiggle on each trace, caused by the first sound from the shot to reach the geophone groups on the spread. The nearest of the groups to the shot point is from 30 to 100 or so meters away, so the first breaks are all later than the uphole break. The farther a group is from the source, the later its first break is recorded, so the farther down the trace it appears. So the line of first breaks on the record slants away from the near trace, the trace from the group nearest to the source. If the spread is single-ended, extending in only one direction from the source, the first breaks will slant down the record from the near trace (Fig. 1-23). If the spread is split, they will slant in two directions. The first breaks for most of the traces arrive by refraction along interfaces near the surface. So

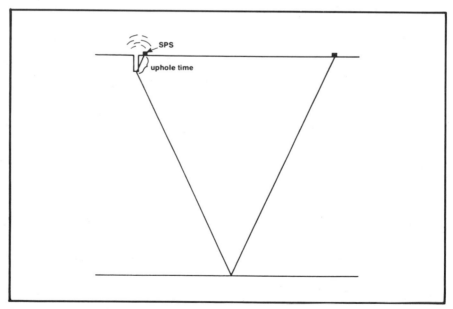

Fig. 1-22 Uphole time measurement

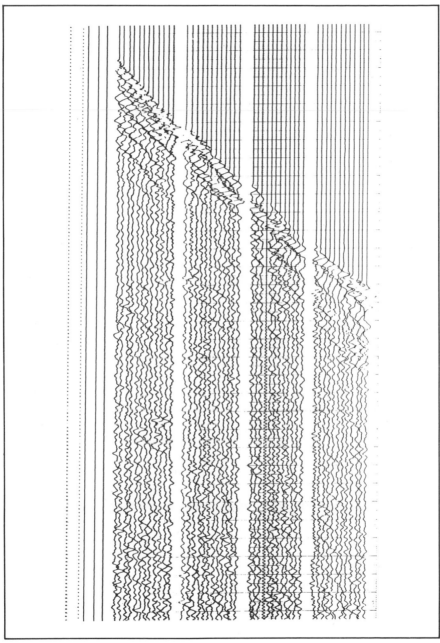

Fig. 1-23 Offshore monitor record   *Credit: Petty-Ray Geophysical*
Division, Geosource Inc.

the first break lineups are normally in straight lines, or segments of straight lines, indicating the velocities of the layers the sound travels along.

After the first breaks are the reflections. They appear as wiggles that line up, from trace to trace, across the record. The energy from these deeper layers is not refracted, but reflected to the geophones. Refraction along those layers would be received by geophones much farther away. The reflections appear on the record as curved alignments of wiggles, bending downward with distance from the source, rather than the straight line segments of the refracted energy. The shallowest reflections curve considerably, and the curvature becomes less for deeper reflections. The curvature is normal moveout.

Seismic Sections

Seismic sections, that is, vertical sections, represent cross sections of the subsurface (Fig. 1-24). They can be made from either 2-D or 3-D shooting. They can be paper prints or displays on computer video screens.

The sections are made up of traces. A trace can be the information from one shot that was received by one geophone group, as on a monitor record. Or it can be the information from a set of traces combined into one. The traces may be displayed in the form of wiggle traces, as on a monitor record. If the wiggles to the right, that is, the peaks, are filled in (in data processing), they are wiggle-VA, wiggle-variable area. Just the fill-in without the wiggles forms VA, variable area, traces. Or the traces may be variable density, VD, with the peaks and troughs (wiggles to the left) represented by narrow bands of alternating dark and light with gradations in between.

The word "trace" came from early earthquake seismographs—a wiggly line was traced with a scriber on carbon-coated paper as the earth shook. The terminology of the wiggles comes from early seismic exploration. Before magnetic tape recording was available, seismic information from a shot was recorded on a piece of photographic paper. It looked like a monitor record. These early records were worked by laying them on a desk with the end representing the surface of the ground to the left and deeper in the ground to the right, just as monitor records are placed for inspection. So the wiggles of the trace were, seen in this position, wiggling up and down (Fig. 1-25). The upward wiggles, looking like a sketch of mountains, were called peaks, and the downward wiggles troughs. The names have stuck, even though sections are often viewed with the surface of the ground at the

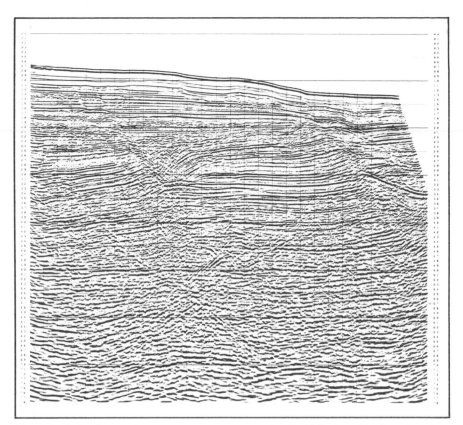

Fig. 1-24 Seismic section   *Credit: Teledyne Exploration Co.*

top (Fig. 1-26), and even though the trace may not be in the form of wiggles.

Traces from a number of shots are combined by CDP stacking. A large number of these stacked traces are placed side by side to make a seismic section. There are usually hundreds of overlapping traces on a section.

The scale of a seismic section is in two different measurements. The horizontal scale is in distance. It represents the length of the seismic line on the ground. And the vertical scale is in time. It represents the amount of time it took the sound to get down to the reflecting surface plus the time to be reflected back up. This is the reflection time, also called two-way time.

A conventional paper print of a section has timing lines, horizontal lines at intervals of typically .01 second, to enable a person to read the

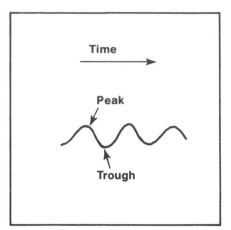

Fig. 1-25 Terminology of wiggles   Fig. 1-26 Vertical trace

reflection times. Shot point numbers are along the top of the section. They can be matched with the same numbers on a map, so data from the section can be plotted on the map. Each section has a header—a label that gives line number, the name of the area, information about the way the line was shot and processed, etc.

A section displayed on a video screen does not need timing lines, as the computer can read any time automatically. It also does not need a header, as the section is called up by line number and shot point numbers. Other header information is available if needed.

The reflections run across the section, but without the normal moveout curvature of reflections on monitor records. The curvature is removed in data processing.

The reflections on a section more or less show the underground rock layers in cross section. With limitations, the rock layers are shaped as the reflections indicate. The limitations of the section in depicting the subsurface will be taken up in later parts of this book.

To use a section, a person usually marks a reflection on it. Then the times to the reflection are read at the shot points. These times are plotted on a map, along with times from other sections, and contoured. The contours show the configuration of the rock layer, from which locations for drilling wells may be deduced.

These sections (and other seismic displays) are discussed in more detail in Chapter 6, Seismic Data Displays, following the description of data processing, which produces the displays. The interpretation of sections is discussed in both Chapter 8 and Chapter 9.

Refraction of Sound

When sound strikes a velocity interface, going from a substance that transmits it at one velocity to a material with a different velocity, only part of the sound is reflected, and the rest continues on its way. However, its path is bent at that point, so it goes in a different direction (Fig. 1-27). This change of direction is refraction. The amount of change depends on the amount of difference in velocities. No difference, no bend. Slight difference, slight bend. Etc.

Also, the bend depends on the angle at which the sound strikes the interface (Fig. 1-28). Perpendicular, no bend. Slight angle, slight bend. Etc.

When the velocity interface is a change to a faster velocity, the bend is in the direction of increasing the angle from the vertical, making the sound go farther away from the source. In the earth, most of the changes are of this type, since compaction tends to make deeper rocks denser and higher velocity. So most seismic refraction, when the sound is traveling downward, bends the sound away from the source. Coming back up it is bent toward the source (Fig. 1-29), with the path more nearly vertical as it approaches the top of the ground. The interfaces on the way up are mostly changes from faster to slower.

Downward-traveling sound that meets an interface in which the change is to a slower velocity rock, will be bent so it is more nearly vertical (Fig. 1-30). In all the situations—fast to slow and slow to fast, traveling downward or upward—the path of the sound is more nearly vertical in the slower material, less vertical in the faster.

The refraction of sound in rock is the same phenomenon as the refraction of light in substances that it can travel through, that is, substances that are transparent to light—air, water, glass, etc.—and that

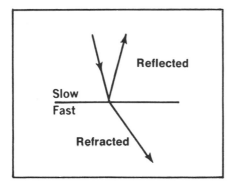

Fig. 1-27 Refraction

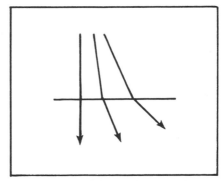

Fig. 1-28 Bend depends on angle

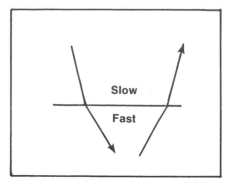

Fig.1-29 Both paths refracted

Fig. 1-30 Fast to slow refraction

makes a stick standing in a pond look bent at the air-water interface (Fig. 1-31).

For any slow to fast rock interface, the sound arriving at an angle leaves at an angle farther from the vertical. So there is some angle of arrival for which the sound continuing through the faster rock goes parallel to the interface (Fig. 1-32). This is the critical angle. Beyond it, reflection does not take place. Instead, sound travels along the interface in the faster material. As it does so, it vibrates the rock, and sound travels up to the surface from each point along its path. The quickest path for sound to reach geophones placed beyond the point of critical angle is down at that angle, along the interface, and up at the critical angle, but in the forward direction. The direct path, spending more

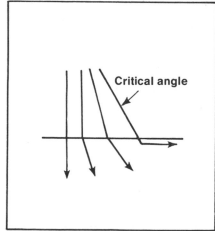

Fig. 1-31 Refraction of light

Fig. 1-32 Refraction along interface

time in the low-velocity material, takes longer to reach the geophone (Fig. 1-33).

Refraction plays a part in all seismic exploration. When sound is reflected, the paths are straight lines only in the oversimplified situation of going through rock of all the same velocity. Where different layers of increasing velocities are encountered, there are bends away from the vertical at all the interfaces (Fig. 1-34). If there are not discrete beds, but only one massive formation, there will be an increase of velocity caused by the deeper parts of the rock being compacted by the weight of the shallower parts. The path of the reflection then will be curved (Fig. 1-35), as though there were an infinite number of interfaces.

If geophones are placed beyond the reflections, where the critical angle causes the returned energy to use a path refracted through a high speed layer, then refraction is used to determine things about the subsurface. But "beyond the reflections" is relative to the depth of the high speed layer. Geophones that are properly placed to receive reflected energy from deep horizons will receive refracted energy from shallow beds (Fig. 1-36). The shallow refraction information is used in reflection exploration to obtain information about near-surface irregu-

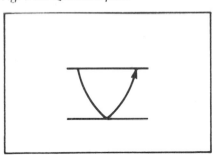

Fig. 1-33 Quickest path

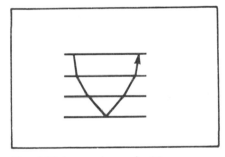

Fig. 1-34 Increasing velocities

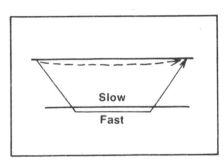

Fig. 1-35 Steadily increasing velocity

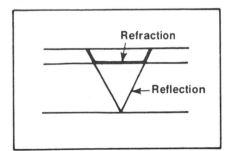

Fig. 1-36 Near-surface refraction

larities. That information is then used to remove the effects of those irregularities from the reflection data.

Geophones placed much farther away, like several miles from the source, are at a proper distance for receiving refraction information from the deep horizons. This is used in the refraction method of seismic exploration. Refraction shooting, the first method of seismic exploration, has been so thoroughly replaced by reflection that refraction is now used only for some special situations. Chapter 10 is devoted to refraction exploration and another different seismic method, shear wave exploration. Otherwise, where seismic exploration is mentioned, it will refer to conventional reflection exploration.

CHAPTER 2

Phenomena

Seismic investigations deal primarily with reflections. Other phenomena add to the picture, or interfere, and must be taken into account. Some phenomena require correction, some must be attenuated, to make the reflections more interpretable.

Low Velocity Layer

The term "weathering" has a meaning in geology and a meaning in geophysics, but the meanings are not the same. In geology, weathering is material that has been broken up (by wind, temperature change, water, plants, animals) and that has not been consolidated. In geophysics the term refers to the material at the surface that has a considerably slower velocity of sound than slightly deeper rocks. Often the change is abrupt, say from 2000 ft/sec to 6000 ft/sec. The geophysical meaning probably originated from the assumption that the slow material at the surface was the geological weathering. Sometimes it is, but another influence can affect the velocity—the level of the water table. Sound travels through air at about 1000 ft/sec and through water at about 5000 ft/sec, so the change from soil with air between the grains to soil with water is abrupt. The level of the water table can change with the season, or with recent rains, etc., so seismic weathering in an area can change from month to month, or even from day to day.

To avoid the confusion with the two meanings of the word, some geophysicists refer to the seismic weathering as the low velocity layer. However, there may be deeper layers with low velocity, so there is another possible confusion. Paul Lyons introduced the term LVL to apply specifically to the low velocity layer at the surface. This expression avoids both chances for confusion. Although weathering and its symbol Wx are in fairly common use for the upper slow material, LVL will be used in this book to avoid confusion.

The LVL is a critical zone in seismic operations. When explosives are used in holes, shots fired within the LVL tend to yield low-frequency data, as the loose dirt doesn't transmit high frequencies well. The long, spread-out wiggles on the traces don't provide as finely detailed information as the closer-together wiggles of high-frequency data, so it is preferable to fire shots below the LVL. Then the higher frequencies are transmitted better, and the uphole time, from shot to surface, includes all of the LVL, so the reflections can be corrected for the time spent in this slow material. Surface sources must transmit this energy through the LVL. And when they are used, LVL thickness must be determined by refraction or by other means.

Normal Moveout

On a monitor record (a record of one shot) or on a CDP gather (a playout of the traces for one common depth point) reflections are curved. The curvature is downward from the shot point, and is concave downward (Fig. 2-1). This curvature is called normal moveout, since it is the normal way reflections behave as the traces get farther (move out) from the shot point, rather than representing dip or any other local phenomenon.

To see why reflections curve, consider the simple situation of a spread in one direction from the shot point, and a single flat reflection. A geophone very near the source will receive energy that traveled almost straight down to the reflecting horizon and almost straight back up. A more distant geophone receives energy that went out at an angle to the reflecting horizon, and up at an angle to the geophone (Fig. 2-2a). So, even though the horizon is flat, the sound traveled a longer path, taking a longer time to get to the geophone. This made the reflection appear farther down the record, so it looked deeper.

Now consider these two energy paths as though they were sticks, hinged at the reflection points. Unfold them and let them hang straight down, to compare their lengths. Suspend them from the midpoints between source and receiver, so they will go through the depth points (Fig. 2-2b). Then they look more like traces showing two-way time, at the points they apply to in the subsurface. Looked at this way, of course the distance down to the reflection on the far trace is farther than it is on the near one. If other traces are added at intermediate distances, the positions of the reflection on them will form a curve (Figs. 2-3a & 2-3b). That curve is the normal moveout curve. Now drawing the same thing again, but for a deeper horizon, the curve will be similar, but flatter, as the path of the sound for the deeper horizon is more vertical than that for the shallower horizon (Figs. 2-4a & 2-4b).

The velocity of sound in the different kinds of rock enters into the picture too, as all the difference between the paths for near and far traces for the shallow horizon is in the shallower, slower material, so the times are longer.

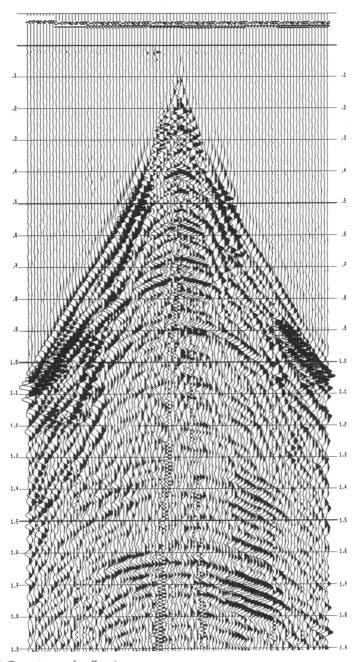

Fig. 2-1 Curvature of reflections
          Credit: Petty-Ray Geophysical Division, Geosource Inc.

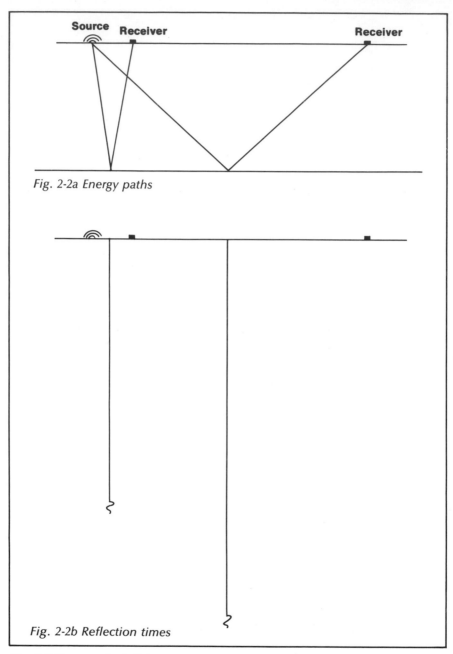

Fig. 2-2a Energy paths

Fig. 2-2b Reflection times

Fig. 2-2 Two receiver distances

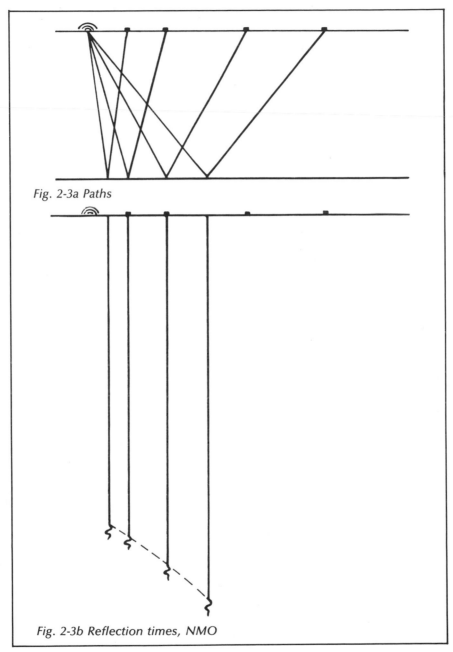

Fig. 2-3a Paths

Fig. 2-3b Reflection times, NMO

Fig. 2-3 Several distances

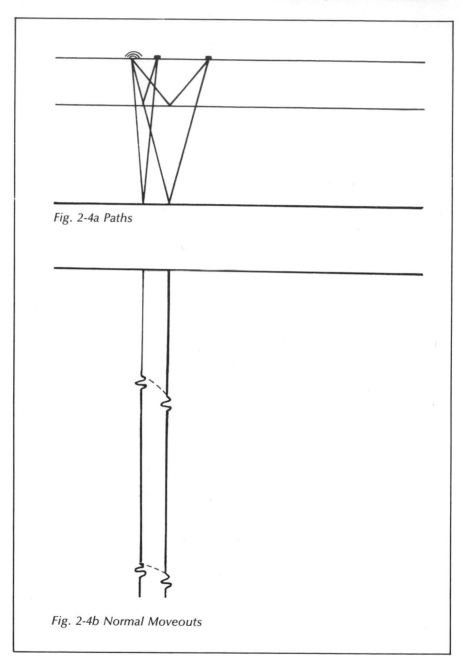

Fig. 2-4a Paths

Fig. 2-4b Normal Moveouts

Fig. 2-4 Two horizons

For brevity, normal moveout is referred to as NMO, or sometimes just moveout. The word "moveout" alone is not clear though, as any other change in time to a reflection on the record, caused by dip, etc., is also moveout, but is not normal moveout.

Multiples

Sound is reflected from places where there are sharp differences in the velocity of sound transmission. A seismic reflection on a section represents sound that goes down to a rock layer of different velocity, is reflected, and comes back up to the top of the ground (Fig. 2-5). However, the base of the LVL has a strong velocity contrast too. So, some of the energy should reasonably be expected to be reflected back downward. It is. And some of that is reflected off the rock layer, up to the surface again (Fig. 2-6). This is a multiple reflection, or just a multiple. The geophones receive reflected energy twice from that rock layer. The second one, the multiple, takes about twice as long as the first one, as it goes most of the same route, but twice. So, on the section it appears to represent another rock layer, below the first one (Fig. 2-7). The section has, instead of information from two horizons, two sets of information from one horizon. The multiple's reflection time read from the section is about twice the time of the first reflection. In this context, the first reflection is called the primary reflection, or just the primary. A measurement using dividers will show the multiple to be about twice as far down the section.

And, if there is enough seismic energy, what's to prevent another bounce from the surface, and another multiple, appearing still deeper

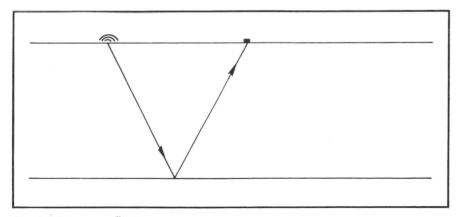

Fig. 2-5 Primary reflection

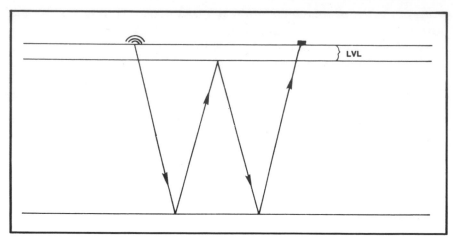

Fig. 2-6 Multiple reflection

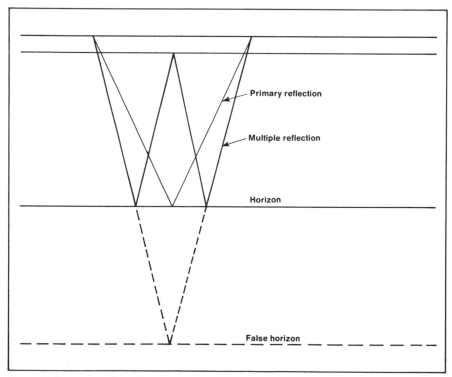

Fig. 2-7 Apparent deep horizon

on the section? Nothing. That multiple will be at about three times the time of the primary, as it traversed the same path three times (Fig. 2-8). And so on, making multiples, getting weaker and weaker.

The top of the ground also has a velocity contrast, and it, too, can reflect sound downward to produce a multiple.

The multiple's time was given as about twice the primary's, rather than exactly twice. This is because the primary goes from shot to horizon to geophones on top of the ground, but the multiple's extra path may be from horizon to base of LVL to horizon. And if the source is an explosive in a shot hole, the primary starts at the depth of the charge, but the multiple goes from base of LVL or top of ground.

Multiples can take several forms, depending on what layers they bounce between and how many times they bounce. Since they diminish as they bounce back and forth, there is a limit to how many times energy can be reflected and still be strong enough to be recorded.

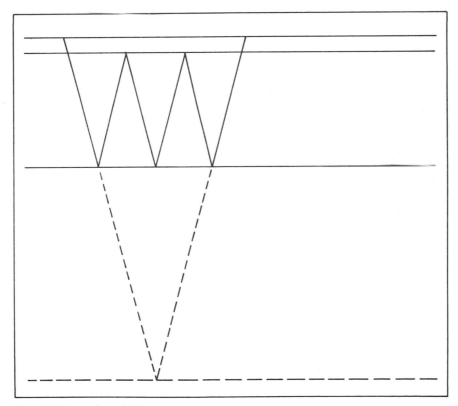

Fig. 2-8 Second multiple

However, in data processing the deeper reflections and weaker energy
are enhanced, and the multiples are unavoidably enhanced with them.

Offshore, at places where the water bottom is hard, multiples are
produced by reflecting surfaces at both top and bottom of the water.
These can both be interfaces with strong velocity contrast. Sound can
bounce back and forth between these two surfaces a number of times
(Figs. 2-9 & 2-10), so the section is dominated by multiples of the water
bottom. This special multiple situation is referred to as reverberation. A
section with strong reverberations is spoken of as ringing or singing,
and looks pretty much the same all the way down. The reflections seen

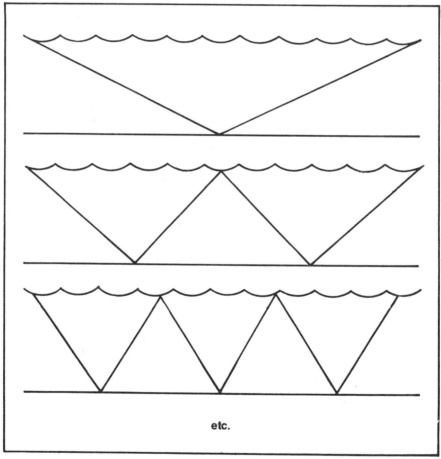

etc.

Fig. 2-9 Water reverberation

are mostly just repetitions of the sea bed. Occasionally there are similar reverberations on land, bouncing between the top of the ground and the base of the LVL.

Multiples can bounce back from any good reflecting surface. Energy may go down to a horizon, up to another rock layer, down to the first one, then up to the surface. This is an interbed multiple (Fig. 2-11), as it does its extra bouncing between two beds of rock.

A reflection that, either on its way down or on its way up (but not both times), bounces between two layers is a pegleg multiple (Fig. 2-12).

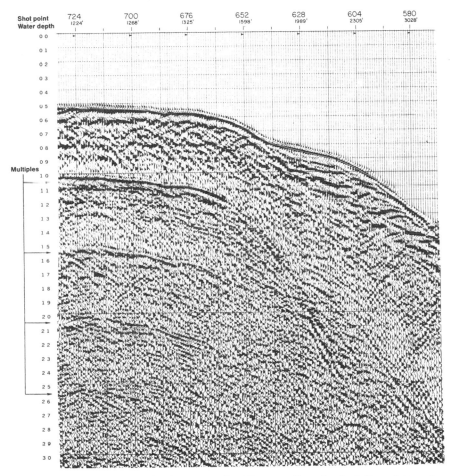

Fig. 2-10 Deep water multiples   Credit: Western Geophysical Co. of America

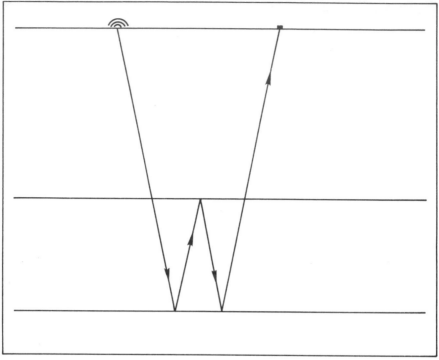

Fig. 2-11 Interbed multiple

The interbed and pegleg multiples do not occur at twice the time of
the primary, but at the time of the primary plus the time spent in the
extra bounce. This description fits all multiples, so:
The reflection time of a multiple is the reflection time of the primary
plus the time spent in the extra bounce or bounces.

In data processing, CDP stack combines traces after correcting for
the normal moveout of primary reflections. This favors the primaries
and weakens multiples. Deconvolution is a process that seeks repeti-
tious events within a specific interval and removes all but the first
occurrence. It also reduces multiples, and is particularly applicable to
sea-bed reverberations. Reverberations on land can be handled in the
same way, but perhaps not as well, because shooting on land does not
normally have the uniform conditions (of elevation, near-surface velocity, etc.) that are normal offshore. Stacking and deconvolution are discussed in Chapter 5, Data Processing.

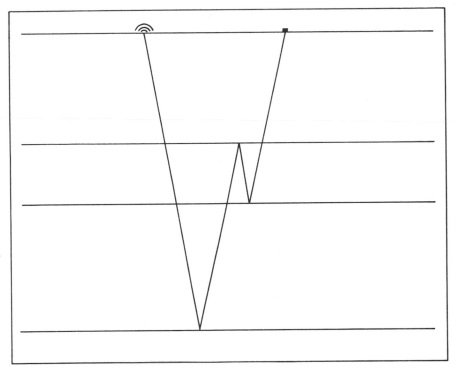

Fig. 2-12 Pegleg multiple

Ghosts

A ghost is a special kind of multiple.

When the source is an explosive in a hole, the shot is usually made below the LVL. Sound from the shot goes out in all directions. Some goes downward and is reflected back by formations. Some, of course, goes upward, and part of that may be reflected back down by the base of the LVL. Then it too can be reflected by the formations (Fig. 2-13). This is a ghost, a multiple that is very close to the primary reflection, one that appears with every reflection on the trace.

The shot usually isn't far below the LVL, so the ghost, reflected from the LVL, isn't far behind the more directly reflected energy. It is often so close behind that the two blend into one complex band of energy (Fig. 2-14). This, in itself, wouldn't be very harmful, but if the shot had been made at a different depth, the reflection and ghost would have

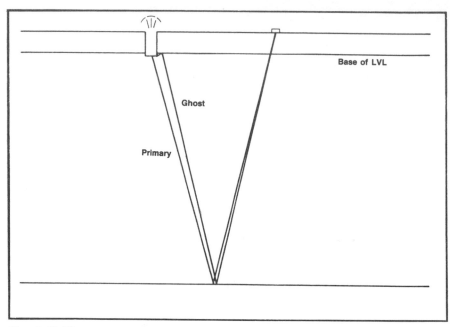

Fig. 2-13 Ghost

combined to make a different-looking band of energy (Fig. 2-15). And a shot fired in another hole may also be different. So the characters of the reflections vary from shot point to shot point, not with changing geology, but with vagaries of shot depth. This can make it very difficult to determine what is going on in the subsurface.

Ghosts may also occur when sound is bounced back from the surface of the ground, rather than from the base of the LVL (Fig. 2-16); or when a surface source is used, the sound may take the extra path from surface to base of LVL to top of ground (Fig. 2-17).

In processing, deconvolution can reduce the repetition of shapes that ghosts produce. This is called de-ghosting.

Bubbles

In offshore work, the most distinctively water-type problem is the formation of bubbles.

An explosion, any explosion, is just a very sudden expansion of something. Usually a solid or liquid becomes a gas, or a gas is released from confinement. In either of these cases, there is, suddenly, a rapidly expanding gas.

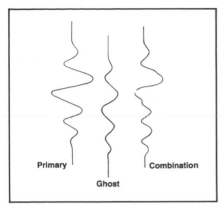

Fig. 2-14 Effect of ghost on primary

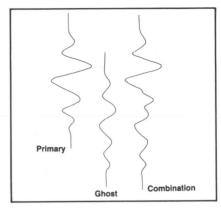

Fig. 2-15 Different hole depth

When the explosion takes place under water, the bubble pushes water ahead of it as it expands. The water immediately ahead of the gas moves at a high speed (Fig. 2-18). After some short period, the expansion is no longer powerful enough to move the water at that speed. But the fast-moving water can't stop immediately. Its momentum causes it to continue to move outward from the site of the explosion. When the momentum is used up, the bubble of gas is overextended, so the weight and pressure of water cause it to contract (Fig. 2-19). The leading water moves inward so rapidly that its momentum causes it to go so far inward that it recompresses the gas. The compressed bubble then expands again (Fig. 2-20). And so on, in diminishing expansions and contractions.

The second expansion is sudden, like the original explosion, although not as strong. To the seismic instruments and on the recording it appears as though a second shot, not quite as strong, was fired. But of course it isn't a convenient time for another shot. Reflections from the first one are still being recorded. So the two sets of reflections, from the first and second expansions, are all mixed together. And third and later expansions and their reflections are also mixed in.

In the various shooting methods used, there are several techniques to reduce the effect of the later expansions relative to the first one, both in special design of the source, and in processing.

Noise

The word "noise", in seismic exploration, is like the word "weed" in gardening. A weed isn't a certain kind of plant. It's just whatever grows that isn't wanted.

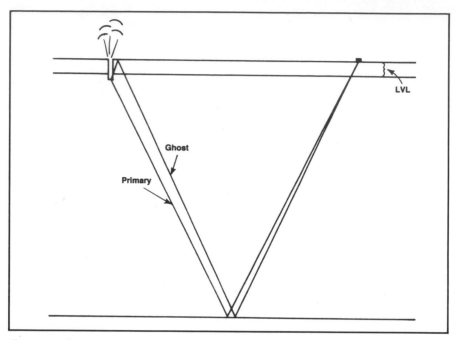

Fig. 2-16 Alternate ghost

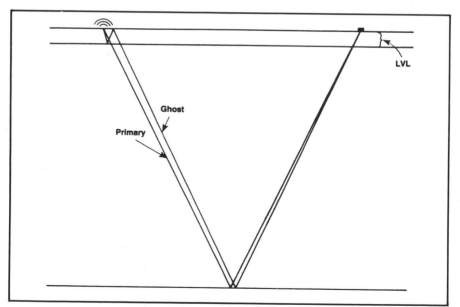

Fig. 2-17 Surface-source ghost

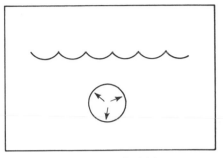

Fig. 2-18 Underwater bubble   Fig. 2-19 Contracting bubble

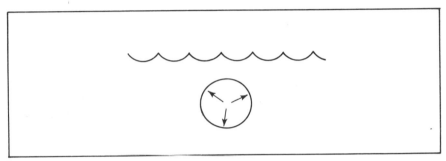

Fig. 2-20 Second expansion

Similarly seismic noise is the sound that isn't wanted. It contains information, but not the kind of information wanted at the time. Generally, the wanted energy is the primary, one-bounce, reflections, with most other energy considered noise. Effort is made to enhance signal, the wanted information, and weaken noise; to increase the signal to noise ratio, or S/N.

Kinds of noise encountered in seismic exploration include:

Multiples—repeats of reflections;

Diffractions—curves formed at sharp breaks of reflecting horizons;

Random lineups on sections;

Outside noise—wind, cattle, boat engines, etc.;

On land

Hole noise—clods, etc., falling after a shot in a hole;

Ghosts—reflections from the base of LVL or top of ground;

Offshore

Bubbles—repeated expansions of an explosion;

Reverberations.

There are different techniques for treating these different types of noise, in field techniques and in data processing. Filtering can reduce the high and low frequencies that are not from reflections. Multiples are attenuated by stacking and deconvolution. Diffractions are diminished by migration.

Wavelet

An explosive seismic source produces a sudden, brief sound containing all frequencies. In going through the earth to a reflecting horizon and back to the surface, and then through the recording instruments, the impulse is stretched out into a wavelet. A wavelet on a trace has one or two peaks and one or two troughs, and extends for about 50 to 100 milliseconds (Fig. 2-21). The peaks and troughs are of varying amplitudes, the highest amplitude occurring about 30 milliseconds after the start of the wavelet, and later wiggles becoming smaller to the end of the visible wavelet. A wavelet in this form is called a minimum-phase wavelet, referring to its having most of its energy near its start. The wavelet is the basic seismic response, the thing that would be recorded if there was only one reflecting horizon in the subsurface.

With just that one velocity interface, the wavelet recorded would have a polarity, that is, a specific wiggle would swing to one side or the other, depending on which way the velocity change went, slow to fast or fast to slow. The wavelet would also have an amplitude depending on the contrast between the velocities (and densities).

In the real earth, there are a great many velocity interfaces, some with large contrast and many with small, some in one direction and some in the other. So the trace recorded is composed of a great number of wavelets, some large and some small in amplitude, some turned one way and some the other, interfering with each other and blending together into a trace in which the individual wavelets are not separately visible (Fig. 2-22).

Another type of wavelet, a zero-phase wavelet, is symmetrical about its highest peak or trough, which is at the time of origin of the wavelet, the velocity interface in the rock (Fig. 2-23). Half of the wavelet is above the interface position on the section, and half below. Obviously, part of the wavelet couldn't be recorded before the seismic time of the interface. So this is not a type of wavelet that is recorded, but one that is produced in processing by re-arranging the recorded data.

Velocity Variation

Seismic velocities are subject to variation, both vertically and horizontally.

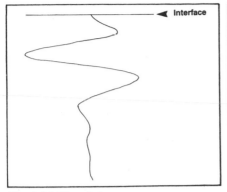

Fig. 2-21 Wavelet

Fig. 2-22 Trace, made up of wavelets

*Drawn from printout by Teledyne
Exploration Co.*

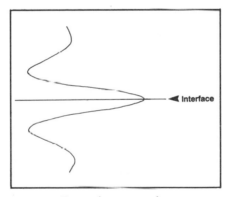

Fig. 2-23 Zero-phase wavelet

*Drawn from printout by
Teledyne Exploration Co.*

The velocity of sound in rock varies vertically with compaction of the rock, and with type of rock in the layers. It varies horizontally with lateral variations in the makeup of the layers. The compaction of the rock is caused by the weight of the rock on top of it; so the deeper the rock, the more it is likely to be compacted, and the higher the velocity tends to be. But changes in the type of rock can go either way. A change from a shale to a carbonate, in the shallow parts of the subsurface, usually produces an increase in velocity, and from a carbonate to a shale, a decrease. Compaction increases with greater weight of over-burden, duration of burial, and compressibility of the rock; that is, how hard and how long it was mashed, and how soft it was. A shale is very compressible, so it increases in velocity with depth fairly rapidly. A carbonate is more resistant, and retains about the same velocity at most depths of interest. Salt is almost completely incompressible. So a salt dome has a velocity of about 14,500 ft/sec all through its height, while the surrounding sediments increase in velocity with depth. They are slower than the salt at shallow depths, and can be faster than it is in the deeper part of the section.

So, in general, velocity increases with depth, but at an uneven rate, and with occasional decreases.

Horizontally, velocity changes with lithologic change and with dip and faulting. If a sand grades into a shale, a velocity change can be expected. And if there is a lot of dip, then at a certain depth and a different location, there will be a different formation with a different velocity. Also the velocity of a formation will change with dip, as its depth of burial and therefore weight of overburden changes. Faulting has the same velocity effects as dip, in putting a formation at a different depth.

If a formation was at one time buried deeply, so it was compressed and became a denser rock, then even if later movement has brought it to a shallower position, it will stay dense and retain its high velocity. Or it may expand a bit, losing some of that velocity. Age of the formation is a rough approximation of this "paleo-velocity". All these variations indicate that velocity is a highly variable factor in seismic exploration. With steep dip or faulting, velocities obtained from a well may not apply a mile away from the well. In other situations, velocities may be consistent for many miles.

Velocity is discussed in greater detail in other parts of this book. The ways of obtaining velocity information from seismic data are described in Chapter 5, Data Processing. Other sources of velocity information and the use of velocity in producing synthetic seismograms are in Chapter 7, Additional Processes. The use of velocity in identifying reflections is described in Chapter 8, Interpretation Procedures.

3

Field Operations

As the first chapter indicated, seismic exploration is not all done by looking at sections and deciding where to drill for oil. Most of the people involved in seismic exploration, most of the time spent on it, and most of its cost are used to obtain the data in the field. This part of it is also the part that feels most like exploration, with travel, weather, hardships, adventure (Fig. 3-1).

Program Planning

A plan of lines to be shot, a seismic program, is made before the crew begins shooting. It consists of lines drawn on a map, so coordinates of the lines can be determined and used to position the lines in the field.

The program may be a grid of lines, particularly in a large new area. It may be detail lines, filling in a prospective area that was discovered with an earlier grid. It may be made to go through well locations, so

Fig. 3-1 Seismic camp in jungle

well data can help identify formations on the seismic sections, and seismic and well data can be better worked together. It may be only a line or so to check the basic idea of a prospect found by field geology or from well logs. Or it may be the close coverage of a 3-D seismic survey made up of many parallel lines.

The program must fit the terrain and other surface features. It may be necessary, economic, or required, to shoot on roads, off roads, in valleys, on ridges, off certain landholdings, around lakes or towns. So planning a seismic survey on land calls for consideration of many factors. Air photos, land-use maps, ownership maps, pipeline maps, etc., may be useful in planning. Offshore programs must take into account such things as sand bars and shipwrecks. And even with very good planning, any program will likely be modified by unexpected obstacles.

The program map is usually large scale, so precise locations can be determined from it. It may show wells, old shot points, surveyors' benchmarks, towns, coastlines, streams, political boundaries. On land, a copy of the map goes to the surveyor on the crew, to establish on the ground the locations to be shot. Offshore, a list of coordinates of ends of lines will accompany the map. The coordinates will be used to guide the boat's radio surveying to put the lines in the proper locations.

Some suggestions for program planning are given in Chapter 12, General Considerations.

Field Geophysics

Seismic information is acquired by field crews. They may be almost anywhere on earth—on cultivated land, in a desert, at sea, in a forest, on a mountain, in a city. The climate may be tropical, temperate, or arctic. About the only limitation is geological—that there should be a sedimentary basin, or a good chance of finding one.

So there is great variety in the conditions on field crews. A seismic crew may be on a rolling boat, remembering their last time in port and living for the next time. A crew may consist of about twenty people living in a small town, driving trucks out each morning, and shooting along the edges of plowed fields. A crew may be hundreds of people, hacking a way through a jungle, carrying equipment on their backs, and setting it up frequently to shoot (Fig. 3-2). A crew may live in a camp in trailers, with separate trailers for sleep, kitchen, dining, recreation, power supply.

In seismic field work, there is much room for improvisation. Conditions on the surface of the ground make it difficult to get around.

Subsurface conditions require various arrangements of sources and receivers. The ingenuity of people in solving problems leads to different kinds of solutions (Fig. 3-3).

In this book, the most common techniques and equipment are described, and also some other ways of handling the problems. Any time there are a number of, say, types of source, someone somewhere will think up and use a new one that neatly solves some special problem. So there will necessarily be field techniques and/or equipment that are not mentioned in this book.

Shooting on land and at sea are so radically different that they will be described separately.

On Land

An arrangement is made to have a seismic crew do the shooting. It may be a company crew, the personnel on it employees of the company wanting the work done. Then the arrangement is an interdepartmental matter of requesting the crew, having it available at the time, charging its expenditures to the proper department by bookkeeping, etc.

Fig. 3-2 Crew on jungle trail　　　　　*Credit: Seiscom Delta United*

Fig. 3-3 Sawing ice to prevent noise from reaching geophones through it
Credit: Geophysical Service Inc.

In most cases though, a crew is hired from a seismic contractor. Seismic contractors range in size from a large multinational company to a single crew operated by the owner of the company. Hiring a crew may involve requesting bids from several contractors, or it may be a matter of getting a crew that is available in the area. A contract specifying equipment, supervision, fees, etc., may be signed. Often a client and a contractor find they work well together, and the contractor, after doing the initial work in an area, may be hired each time the oil company needs more shooting in that area.

The crew is given an assignment map, showing the planned program. On a contract crew, the client may plan to have a representative in the field when shooting begins, to see what record quality is obtained, and what problems there may be. The client's representative may also visit from time to time during the progress of the shooting, or may stay on the crew for the entire shooting period.

The contract company usually has a field supervisor, who visits the crew at the start, and makes frequent visits during the shooting. This supervisor is the main liaison with the client while the work is going on, and may be supervising several crews for various clients at the same time.

Permitting

In the United States, where most land and the minerals under it are owned by individuals, permission to shoot must be obtained from the individuals involved.

A crew has one or more permit men, permit agents, to obtain the permission. Before shooting can begin, they investigate the ownership of both land and minerals. The land may be owned by one person or company, the minerals owned by another, the minerals leased to yet another, and another may be renting the surface for farming or grazing. The investigation usually starts with records in the county courthouse, but is supplemented by inquiring around.

To the individuals or organizations involved, the permit agent describes where lines are planned, and asks if they can be shot. There is usually a fee of so much per mile to be paid to the owner or tenant for permission to shoot. The fee varies with the part of the country, largely depending on how much oil exploration has been done in the area. The person occupying the property may give conditions for the shooting, like going along the edge of a field, leaving gates closed or open, etc., and may give useful information on how to get in to places, where water for drilling can be picked up, etc. (Fig. 3-4).

After the crew has shot on the land, there may be a second visit to the owner, to see if there are complaints, and to pay for any unplanned damage, like ruts formed when a truck was stuck.

Surveying

A surveyor, using the locations on the program map, finds where those points are on the ground, marks them for other members of the crew to find, marks routes through gates, etc., to those points, and determines their exact locations and elevations (Fig. 3-5).

There are several types of survey instrument. A transit and a theodolite each has a telescope with a means of measuring angles horizontally and vertically. The instrument is mounted on a tripod, and is used to sight on a rod, a board held upright by an assistant called a rodman (Fig. 3-6). The rod has units of length marked on it in some easily-distinguished pattern so they can be read from a distance. The marks indicate height above the base of the rod. Notes are made of angles and heights, so locations and elevations can be calculated later.

An alidade is another surveying instrument that has a telescope to sight on a rod. The alidade has a flat base, with straightedges along the sides of the base. It is not mounted on the tripod, but is moved about on a plane table, a flat board that is mounted on the tripod. A map is

STAY OUTTA THE COTTON, STAY OFF THE ROADS IF'N THE'RE MUDDY, STAY OFF THE ROADS IF'N THE'RE DUSTY, DON'T CUT NO FENCES. CLOSE ALL GATES, PLUG ALL HOLES, AND THAT'LL BE $50 PER HOLE AND $10 PER LOAD OF WATER — WRITE THAT DOWN SON!

PERMIT

Fig. 3-4   *Credit: J.C. Knight*

drawn on a plane table sheet, a sheet of paper or film secured to the plane table, as the readings are taken. The alidade is positioned at a point on the map and the direction sighted to the rod is penciled on the plane table sheet along the straightedge.

An electronic distance measuring device, EDM, sends radiation to a reflector on the rod, receives it back, and measures the time taken. It performs a calculation to convert the two-way time to one-way distance, and displays the distance on a small screen. The display is immediate, and the distances are accurate. A commonly used surveying

Fig. 3-5 Surveyor   Credit: *Petty-Ray Geophysical*
Division, Geosource Inc.

setup consists of a radar distance-measuring device mounted above a transit or theodolite on the same tripod. Then horizontal and vertical angles can be read from one instrument and distances from the other. A combination instrument can also record readings without their having to be read by the surveyor. The readings are later put into a computer which calculates locations and elevations of the points surveyed.

The survey is run from a known location and elevation at a government benchmark, an abandoned well, etc., by the surveyor with the surveying instrument, and the rodman with the rod. They usually take turns moving forward, with one of them at the last surveyed point and the other at another, not yet surveyed, location. As a check of accuracy, the survey is normally run in loops, that is, extended around to meet earlier locations. This checks on whether the new location and eleva-

Fig. 3-6 Rodman   Credit: Seismograph Service Corporation

tion are the same as the old, or rather, that any disagreement, a mis-tie, is within acceptable limits.

The surveyor and rodman determine starting points and directions of lines. Then layout people place stakes, flagging, metal tags, whatever shows up and holds up well in the area, at shot points and geophone group locations, to mark those points for the other units of the crew. The surveyor and rodman survey those points to obtain accurate locations and elevations. They may also place permanent benchmarks near key points like ends of lines and line intersections, so locations can be found later. And if another seismic program is conducted later in the area, the benchmarks help the surveyor tie the new lines to the old lines.

Some seismic programs may need several surveyors, but even so, surveying is a rather solitary occupation. The surveyor is usually ahead of the rest of the crew, each surveyor accompanied by a rodman. And the rodman is, much of the time, just a small figure seen through the instrument telescope. So the surveyor tends to mumble while sighting, things like "Get out from behind that tree" or "Hold the rod straight, you're leaning."

The surveyor must also be self-reliant, like the surveyor in a cold climate, who told me he'd never had his pickup stuck in the snow, just slowed down. It could always go as fast as he could shovel a path.

Energy Source

Seismic exploration uses sound produced artificially at or near the surface of the earth. There are a great many ways that the sound can be produced. People have exercised their imaginations and inventiveness through the years to develop different ways of producing the sounds, to gain one quality or another in investigating the subsurface. So the ways seismic energy is produced are many and varied, strange and wonderful. An explosive detonated in a hole in the ground is a particularly good source. A controlled vibration imparted to the earth can be used in cities without disturbing the inhabitants. Explosives to be fired on stakes above ground are readily portable. And so on.

Shot Hole

For much of the history of seismic exploration, almost the only type of energy source used was an explosive in a hole in the ground. This is still one of the most common types for shooting on land.

When a charge is to be fired in a hole, the hole is normally drilled by a rotary drill, which is like a miniature oil well drill. A seismic rotary drill is usually mounted on a truck (Fig. 3-7).

The drilling is done by a bit on the bottom end of a piece of drill stem, a heavy steel pipe threaded on both ends. The drill stem is held

Fig. 3-7 Rotary drill   Credit: Petty-Ray Geophysical Division, Geosource Inc.

upright by a mast on the drill. A truck-mounted drill travels with its mast laid down, and the mast is raised to the vertical position for drilling. The drill stem is turned by a rotary table, and the turning bit breaks the rock. A drag bit has a fishtail tip that cuts when it is turned, in the way that you might deepen a hole in wood by poking a screwdriver in and turning it. It works well on fairly soft rock. For harder rock, a rock bit is used. It has three cones or several cylindrical rollers, either being free to turn and having teeth. Rotation of the drill stem turns the cones or rollers, so the teeth chip away at the rock.

With either drag or rock bit, a fluid—mud, water, or air—is forced down through the inside of the drill stem by a pump or air compressor. The fluid goes out through holes in the bit, and up around the drill stem, carrying the cuttings—chips of rock—up with it.

A truck-mounted drill that uses water or mud carries a metal slush pit. If mud is to be used, it is made by adding commercially prepared, powdered, dry drilling "mud" to the water in the slush pit. The mud partially seals the sides of the hole, and being heavier than water helps it carry the cuttings out. The fluid coming out of the hole goes into the pit, where the cuttings can be shoveled out, and the mud or water picked up by a suction hose and pumped back down the drill stem. A driller usually has one helper, who also drives a water truck to keep the drill supplied with water. The drill also has a pulldown, a means for putting some of the weight of the truck on the drill stem to press the bit against the rock.

When the drill stem has reached as far down as it can, another length of drill stem is added at the top, and so on as needed to drill to the desired depth. Holes are usually drilled from 20 to more than 200 ft. deep, most of them in the range of 60 to 100 ft. After the hole is drilled, a winch pulls the drill stem up so it can be unscrewed a segment at a time.

An air drill is similar to a drill that uses water, but doesn't need a supply of water or a water truck. It just uses an air compressor to pump air down the hole, while the bit is turning in the hole. The air blows the cuttings out. It's dusty around there. Some drill trucks are equipped with both a pump for mud or water and an air compressor, so they can be used whichever way is best for the area.

An auger drill doesn't use water or air. Its drill stem is threaded all along its length, so it works the cuttings out along the threads like a wood-drill does (Fig. 3-8).

An air-hammer drill uses compressed air to drive the drill stem so the bit hits the bottom hard. Then the drill stem makes a quarter turn, and the bit hits again, and so on. This chips at the rock. It is particularly good for hard rock.

Fig. 3-8a Drill

Fig. 3-8b Closeup of drilling

Fig. 3-8 Auger drilling   *Credit: Western Plains Services*

For areas that cannot be reached by truck, the drill can be in several pieces, so each piece can be carried by helicopter. They are carried one at a time, lowered to the ground, and assembled by hand. The assembly may be just a matter of connecting hydraulic hoses. After one hole is drilled, the whole thing is taken apart and the parts flown to the next location. A variant is a helicopter-portable drill with also a helicopter-portable chassis, so after the drill is assembled it can be driven from one shot point to another.

Where not even a helicopter is practical, very light drills carried by people can be used. If the drilling in the area is easy, a pumper drill, little more than a water pump with some light drill stem, can be carried down jungle trails. No bit is used. Water is just pumped down the drill stem to wash soft earth out (Fig. 3-9). Or in areas of loose dune sand, a puffer drill with an air compressor instead of a water pump will blow air down the drill stem, blowing sand out of the way. For more difficult

Fig. 3-9 Pumper drill. Man on top is the 'pulldown'

Credit: Seiscom Delta United

drilling, a small rotary drill that breaks down into a number of packages can be hand-carried over trails and assembled at the shot points.

Any type of drill that is used in portable form, can of course also be made larger and sturdier, and mounted on a truck for situations in which the type of drill is effective, but portability is not necessary.

The Shot

After the hole is drilled, it is usually loaded with explosive by the drillers, to be fired when the recording people arrive at that shot point. For use in holes, explosives are packaged in one-piece plastic tubes, threaded on both ends so tubes can be screwed together to make up larger charges. One tube containing a five-pound charge is about two and a half feet long, and several may be assembled to make a single long charge. An electric blasting cap is inserted in a special hole built into a tube, and the cap wire looped around the tube. The charge is lowered into the hole by the wire on the cap, if the hole is unobstructed. But if the sides of the hole are so rough that the charge would catch on irregularities, then loading poles must be used to get it down. They are lengths of pole with connectors on the ends so as many as necessary can be hooked together. A shot-hole anchor with upward-pointing springy prongs may be attached to the charge to hold it down, if there is water in the hole so the charge would float up.

Precautions are taken to keep the cap wire from being pulled up. The hole is covered. To plug a hole before it is shot, it is usually enlarged about an arm's length down it. Then a hole plug is put down on the shelf created by the enlargement. If there are rocks of the right size handy, one of them may be used for a hole plug. Usually a metal plug descriptively called a tin hat hole cover is used. Then dirt is shoveled in to fill the hole to the top.

Later, when the shooters and recorders arrive to fire the explosive, the cap wire is retrieved and attached to recording equipment, and the explosive fired by a signal from the equipment. Mud, rocks, cap wire may spray into the air in a brief fountain (Fig. 3-10). After use, the crew permanently plugs the hole.

It would seem that the best way to plug a hole would be to shovel back all the cuttings that came out of it. But that procedure isn't reliable. Some of the dirt would be likely to lodge somewhere down the hole, and it would be completely filled only from there up. Then, maybe in a rainy time later, the clogged dirt would loosen, and there would again be an open hole. The crew may fill the hole with bentonite, a special mud that expands when it gets wet, and put cement above

Fig. 3-10 The Shot   *Credit: Seismograph Service Corporation*

that. Or a hole plug may be put about 20 feet down the hole, and the
hole filled with cement above that. In either case, the last foot or so
may be filled with dirt. The method used is often dictated by state
regulations.

For many years, the explosive that was detonated in shot holes was
dynamite. Dynamite is sensitive, and has to be handled very carefully.
The active part of dynamite is nitroglycerine. Explosives are still fired in
holes in many seismic operations, but actual dynamite may not be the

explosive used. Even though crews may still refer to their explosive as dynamite, it may be based on ammonium nitrate (a common fertilizer for plants) or some other explosive. There is even an explosive that consists of two components, a solid and a liquid, which must be mixed before it can be activated. The separate components can be shipped safely and mixed only when the crew is ready to use them.

The newer explosives have major advantages over dynamite. They are less likely to be set off by accident, and handling them does not cause headaches. Believe me, a dynamite headache, from the nitro-glycerine's causing blood vessels to expand, isn't any fun.

Another form of an explosive used in a hole is a delay charge composed of explosive cord coiled around a stick, or other means of slowing the effective rate of detonation down the length of the charge. That velocity is adjusted to fairly well match the speed of sound in the surrounding earth. The detonation is initiated near the top of the charge, and spending time going around and around the stick, keeps pace with the force of the explosion traveling through the rock. The energy traveling downward from all parts of the charge leaves the bottom of the charge at the same time. This has the effect of directing most of the force downward to yield seismic information.

Some seismic explosives are in the form of shaped charges. If a solid explosive has an indentation in its surface, then the force of the explosion from the inner surface of the indentation forms a concentration of force, like the focusing of light. The shaped seismic charges have the indentation in the bottom, to direct more of the force of the explosion downward.

Pattern Holes

The discrimination against horizontally traveling noise that is achieved by a geophone group spread out over a horizontal distance of a hundred feet or so, can also be obtained by firing a number of explosive charges in a horizontal line. The pattern may also be spread out to the sides for other noise-cancellation effects.

Several shot holes are drilled in a pattern, and shots fired in them simultaneously (Fig. 3-11). The holes are usually shallower than a single hole would be, and the charges smaller than would be fired in a single hole. The holes may total about the same number of feet as the single hole, and have about the same amount of explosive divided up among them. The shallower drilling may be easier, if it does not reach as hard a formation, but it requires setting up the drill more times. The method is especially applicable to portable operations, where a light-weight,

Fig. 3-11 Pattern shot   *Credit: Seismograph Service Corporation*

light-duty drill can be carried more readily. A disadvantage is that the shots will not be below the LVL.

Surface Explosives

In some areas where portability is critical, shot holes may be dispensed with, to eliminate the need to carry a drill. Then small charges are fired near the surface. Often the charges are in sacks, hung over stakes for air shooting. A fire-retardant chemical is included in these explosives, so they won't ignite dry grass, etc. The sacks are put about 18 inches above the ground, and a number of them are detonated as a pattern charge. The detonation proceeds from one sack to another by an explosive cord strung between them. The shots, fired nearly simultaneously, strike the ground somewhat as a flat force. They have little effect on plants on the surface. This system requires having an area that is isolated, so the sound of the explosion does not disturb people.

Explosive cord is made of an explosive in a woven sleeve, and covered with a plastic coating. It looks about like plastic clothesline. It is used alone in another application of explosives that does not require shot holes. Usually, a plow digs a trench, lays the cord in it, and replaces the dirt, all in one trip. The plow may be pulled by a tractor, or installed on one. Where plowing is impractical, the cord can be simply unreeled onto the ground, although much of its energy will be lost into the air. A special advantage of using cord is that the charge is spread out horizontally to give a noise-cancelling effect like that of pattern holes or a geophone group. For use in forests or other areas of potential fire hazard, the cord has a fire retardant outer layer.

Weight Drop

The weight drop seismic source uses a heavy weight, lifted about ten feet, then dropped on the ground (Fig. 3-12).

The weight is flat, and is dropped so it will land flat. Immediately after the weight hits the ground, hoisting it back up begins. This raises the weight as it bounces, so it won't strike the earth a second time and create a second seismic impulse, with a second set of reflections. Also, it is then quickly ready for the next drop.

Weight drops are used largely in desert areas, where many individual initiations of energy over a large area may be needed to get usable seismic data. Dry, loose sand and soil transmit sound very poorly, but the weight strikes with about the right force to be propagated well by

Fig. 3-12 Weight drop unit

Credit: Petty-Ray Geophysical Division, Geosource Inc.

the desert soil. It is also an advantage to not have to transport an explosive to such remote areas.

A weight drop truck usually has the weight at the back. Light chains dangle around the weight, to warn people. The weight itself may be a piece of battleship armor plate.

A variant of the weight drop for remote areas is a net made of chain dropped from a helicopter. As you may know from dropping similar things, the mass drops with very little rebound.

Vibrator

Vibrators are the most popular type of seismic energy initiators on land in areas accessible by truck. A vibrator produces seismic energy by vibrating a weight up and down. The weight is on a pad that is held in tight contact with the ground, so the vibration is propagated into the earth.

The vibrator is usually mounted on a truck, and some of the weight of the truck is used to hold the pad firmly against the ground (Fig. 3-13). The weight vibrated may typically be around two tons.

The vibrator goes through a sweep, from high frequencies to low (a downsweep), or low to high (an upsweep), in the course of, normally, 7

Fig. 3-13 Vibrator in action

Credit: Geophysical Service Inc. (Photo by J. Funk)

to 15 seconds (Fig. 3-14). Later, in processing, the recorded mixture of initiations and reflections is sorted out so it is as though a single shot had been fired.

The vibrator causes very little disturbance to the surroundings, so it is often used in areas where another type of source might cause damage, like along highways or in cities. It has even been used to record in a tunnel through a mountain.

Land Air Gun

Air guns were originally developed for use in offshore exploration. An air compressor forces air into an underwater chamber, and the air is released from the chamber with explosive force at a depth of about 30 feet in the water. But air guns for use on land have been developed (Fig. 3-15). The way they are made to work on land is by duplicating the conditions offshore. To put the gun underwater, a container of water is used, with the gun placed in it, and the container pressed against the ground. To achieve the effect of being 30 feet down in the water, compressed air is pumped into the water container. Then, with the air gun in water under pressure, compressed air is forced into the gun and then released. The firing imparts seismic energy to the earth much like the way it is put into the sea bed. But the water container is small, and immediately after firing it is vented, so the expanding air escapes before a second bubble expansion can occur. So there is not a repeated bubble problem like there is offshore.

Geophone

A geophone is a device for detecting sounds from the earth— "geo", earth; "phone", sound.

The type of geophone generally used in seismic exploration on land

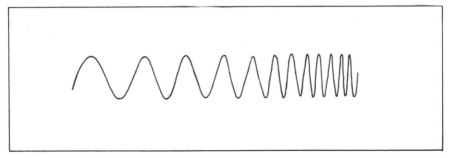

Fig. 3-14 Vibrator sweep

Fig. 3-15 Land air gun   Credit: Bolt Technology Corporation

consists of a powerful permanent magnet and a coil of wire around the magnet, all in a rugged case (Fig. 3-16). The magnet is attached rigidly to the case, which also serves as a retainer for the magnetic field, like the familiar use of a piece of iron placed across the poles of a horseshoe magnet. The coil, made of fine copper wire wound many times around, is connected to the case by a spring. The geophone is placed in firm contact with the ground. Then any shaking of the ground will shake the case and magnet. The coil, suspended on its spring, will not move as quickly. So the magnet moves up and down past the coil. Relative movement between magnet and coil generates electric current, on this small scale, just as it does in the dynamos that power cities. The ends of the coil are connected to a "pigtail", a pair of wires extending from the case, to be connected through the cable to the recording equipment. The bits of current generated by the movements of the magnet make up the signal from the geophone. They are in proportion to the speed the magnet moves past the coil, so this type of detector is called a velocity geophone.

Geophones are set out by jug hustlers, juggies, who put the string of geophones for a group on the ground at set distances from each other. The geophones that make up one group are pre-connected to each other by wire. That mass of wire and geophones is usually carried

Fig. 3-16 Geophone with spike   Credit: GeoSpace Corp.

strung on a clip like a big safety pin, by a person who pulls off loops of wire while walking (Fig. 3-17).

Good contact with the ground is necessary. Some geophones have spikes on the bottom, to be stabbed into the ground. For frozen ground, they may be flat-bottomed, and need to be pressed into place, or grass scraped aside. Placing the geophones on the ground is "planting" the geophones. A good plant is one in which the geophone is upright and the contact is close enough to allow vibration to be transmitted well from the earth to the geophone.

When all geophones are in position, and the groups connected to the main cable or distributive units, an instrument is used to determine that all the groups are connected, none in reverse, no breaks in the cable. Then, or after corrections or repairs have been made, the spread is ready to be shot.

Cable

On most land crews, geophones are spread out in a line, connected by a cable to the recording equipment. The data from each geophone group is transmitted along one pair of wires, so the cable has as many pairs of wires as there are geophone groups: 24, 48, etc. At intervals, the cable has takeouts, pairs of wires to be connected to the groups of geophones. The cable is carried "horsecollar" fashion, looped over a person's neck and/or shoulder.

Voice communication between the instrument truck and the people

Fig. 3-17 String of geophones

*Credit: Petty-Ray Geophysical Division,
Geosource Inc.*

out along the cable is by telephone, using wires in the cable or special
telephone lines that are laid out when setting up, or by radio.

Seismic cables get rougher treatment than most electrical cables.
They are put on all kinds of ground, laid down and picked up repeatedly, sometimes dragged across the ground, chewed by foxes, walked
on by cattle, run over by trucks. So wires in the cable are likely to break.
Seismic crews become adept at locating and repairing cable breaks.

If the cable gets wet, then at places where the insulation on some
wires is cracked or broken, current can be conducted by the moisture.
This connects different pairs, resulting in crossfeed, the blending of
information among different pairs of wires. To reduce this problem, a
cable, especially an old one, must be dried out occasionally. When

shooting on roads, there may be power lines nearby and parallel to the cable. This makes the cable pick up the 50 or 60 cycles from the power lines by induction.

Distributive System

Some shooting, for instance 3-D, uses large numbers of geophone groups. And some recording equipment has over 1000 channels. A conventional cable to transmit 1000 channels of information with a separate pair of wires for each channel would be thick, heavy, unwieldy, and almost impossible to keep in repair. A distributive system is a way of handling that information with much lighter cable. In such a system, a small unit, about a cubic foot in size, transmits data from several groups of geophones to the main recording equipment. The data is broken into bits and sent on a single pair of wires to the main equipment. The unit can also re-transmit data from farther units. So the cable consists of one main pair of wires and the separate pairs of wires to carry information from the groups to the distributive units. This also largely eliminates the problem of crossfeed. With just the one pair of wires sending data at the moment, there isn't any other pair for its data to get mixed up with.

Another type of distributive system uses glass fibers to send information in the form of light, instead of wires carrying electricity (Fig. 3-18). The main cable has two strands of fibers all the way down its length, and short lengths of wire to connect to the groups. This system has the same advantage in eliminating crossfeed that the two-wire system does, and also is immune to induction from power lines. Induction is an electrical phenomenon, and doesn't affect light.

Another method uses radio telemetry, transmitting the data from one or several groups of geophones to the recording equipment by radio. This can work where the terrain is so rough that it isn't practical to run any kind of cable from the groups to the recording unit. The transmitters and strings of geophones might be carried by people or deposited by helicopter. Transmitters with geophone groups can be set out in jungles, reducing the amount of trail to be cut and making difficult access easier by eliminating the laying out of cable continuously along the line.

The recording equipment for radio telemetry does not have to be near the geophones, but does need to be more or less in line of sight of them. One way this requirement has been satisfied is to have the recording instruments in an airplane circling over the vicinity of the spread.

Fig. 3-18 Data transmitting unit

Credit: Petty-Ray Geophysical Division,
Geosource Inc.

Recording

A seismic source produces sound that goes into the earth, is reflected from rock layers, and arrives at the surface of the earth. The geophones convert some of that sound into electrical energy, which the cable conveys to the recording equipment.

The recording equipment is a big, high-quality tape recorder. The cabin or other space containing it is called the doghouse (Fig. 3-19). The doghouse can be mounted on a truck, tractor, marshbuggy, airboat, or other kind of vehicle. Or it can be designed to be carried by a helicopter, looking like a toy on a string, to be set down near the spread to be shot (Fig. 3-20). Or the recording equipment can be made so it breaks down into units that can be carried by people and set out in the open or in a tent.

The equipment is usually connected with the source of the sound by wire or radio, so a signal from the recording equipment turns on the

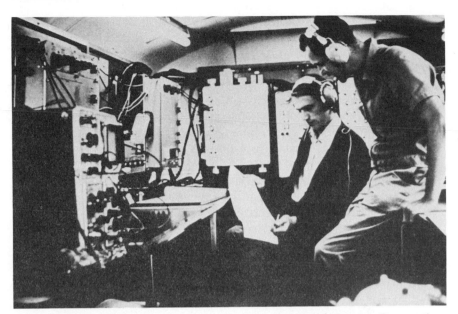

Fig. 3-19 Inside doghouse   *Credit: Seismograph Service Corporation*

source— fires the charge, starts the vibrator, or whatever. The equipment has computer-like panels with the usual indicator lights, numbers, dials, knobs. On one panel are two tape reels, with tape winding from one to the other by way of a number of tension rollers.

The person who runs the equipment is called the observer or operator. When a shot, pop, etc. is fired, the tape reels turn for a few seconds, typically 5 or 6, recording the data. Then the tape stops until the next shot. With an explosive in a hole, the next shot does not occur until the shooter has gone to that hole, connected the cap wire, and made sure no one is close to the hole. With some other sources, the operation is more continuous, with energy initiated maybe every few seconds.

Forms are filled out as a log of times of shots, problems encountered, etc. At intervals, a monitor record is played back from the tape onto paper, and a spot check made on it for problems.

After a number of shots, the takeup reel is full of tape, a new reel of unrecorded tape is put on, and the empty reel put in the takeup position. The recorded reel is placed in a holder and sealed against dust, moisture, etc. It is packed with others for shipment to a processing center, at the next convenient time.

As shooting proceeds, there is a continual change of cable and geophones, segments of cable from behind being brought around and connected ahead of the shooting. This moving around is characteristic of common depth point shooting, which is therefore sometimes called, from a field person's point of view, roll-along.

One particularly portable type of recording equipment is a unit that records the output of one geophone group. It is in a small box that can be hand-carried. It is connected to the geophone group and records the output of only that group on a tape cassette. So there must be one unit at each group. When the source is to be initiated, the units are turned on by radio from another type of portable unit. Only one of the second type is needed for the entire spread. At the end of the day, the cassettes can be gathered and transcribed in the crew office to the larger reel of tape that is standard in seismic recording.

Offshore

A seismic crew at sea operates in a very different way from a crew on land. Rather than consisting of units operating separately for surveying, cable layout, shooting, recording, etc., the crew is all together on one

Fig. 3-20 Doghouse placed by helicopter

Credit: Seismograph Service Corporation

boat. There aren't the discomforts of plodding through swamps, jungles, and snowstorms. But the boat isn't steady like the ground. It rolls, pitches, and yaws 24 hours a day. And the work goes on 24 hours a day, too, so repairs and maintenance have to be done between lines or while shooting. The shooting is much faster, and great quantities of data are collected.

Offshore, the leases, concessions, or production sharing contract areas are obtained from governments, and permission to shoot is included in the arrangement. In addition, specific programs may require approval before shooting. A government representative may ride the boat throughout the entire project.

Vessel

A seismic boat for normal offshore exploration is from about 150 to 200 or more feet long and about 40 to 50 feet wide. It tends to be shallow draft, around 10 to 15 feet, so it can work in close to shore. It is equipped to stay at sea, shooting, for a month to several months without supplies from land (Fig. 3-21). The crew consists of a ship's crew and a seismic crew. Both crews are double staffed to keep the operation going day and night. The crews usually alternate in six-hour shifts, with the breaks coinciding with meal times, so the people going on

Fig. 3-21 Seismic boat   *Credit: Geosource Marine*

duty can eat before they start, and those going off can eat after work.

The quarters are reasonably good, with from one to four people in a cabin. The crews may work 28 days in a row, followed by 14 or 28 days off, or they may work through a shooting program and then have some time off. Most boats have a recreation room, with a television set and maybe a videocassette recorder. The food is usually excellent and plentiful. In addition to regular meals, there is usually an arrangement so that individuals can snack between meals, provided they leave the place clean afterward.

Most seismic boats are designed to have room for all the original equipment. But during the course of several seismic programs temporarily needed gear is often added and rarely removed, so the boat tends to get more cramped. The additions are often at deck level or higher, so they raise the center of gravity, which makes the boats tend to roll constantly.

Things that would, on other vessels, be solely the skipper's responsibility are, on seismic boats, often controlled to fit seismic requirements. Navigation becomes a matter of staying precisely on a line, speed dictated by recycling time of the source and frequency of pops so that a shot is completely recorded before the next one is fired.

An intercom system allows communication between recording room, wheelhouse, and back deck. And television screens are often put at key places around the vessel, to allow people to monitor activities wherever they may be. A TV camera may be on the back deck, scanning whatever layout or repair activity may be going on there. Screens showing what that camera sees may be in the recording cabin, the wheelhouse, and the dining and lounge areas. Similarly, data on lines shot and ship's course may be shown (Fig. 3-22).

Offshore Surveying

The surveying for an offshore crew can't be done ahead of time, as on land, so it is handled by some system that can survey in real time, as the shots are made. The terms "navigation" and "positioning" are often used to refer to offshore surveying. Most of the systems use radio, some of them by receiving broadcasts from earth satellites, and others by measuring distances from transmitters at known locations on land.

The survey system, given the first and last points of a line to be shot, provides instructions for guiding the boat on an approach to the line and along the line, and also signals when shots are to be fired. The directional information can be displayed in the navigation room, and also in the wheelhouse, so navigation for the person at the wheel is just a matter of keeping a pointer on a line on the screen, or setting headings that appear on the screen. Directions are also given for the approach to the next line, in a wide swing so the cable will be out straight behind the boat when the line is started.

By Satellite

Satellite systems use orbiting artificial earth satellites. The satellites are in polar orbit, so that as the earth turns below them, they pass over all parts of it. One system is called TRANSIT, Navy Navigation Satellite System, or NNSS. A satellite broadcasts two stable frequencies, and includes in the tones a time signal and data on the location of the orbit.

The satellite signals are picked up by a specialized radio receiver on the boat. Although the tones have a steady pitch, they are received at a higher pitch as the satellite approaches the boat, and a lower pitch as it leaves. This is the doppler effect, which applies to any frequencies from any source that is moving with respect to the receiver. The change is more pronounced for a pass near the boat, and is smoother for a farther pass. A computer with the receiver on the boat uses frequency changes, orbit location, and timing to determine the boat's position.

Fig. 3-22 Video displays on seismic boat

Successive satellite passes can be as much as several hours apart. With the satellites in polar orbits, going north and south, passes are closer together near the poles than near the equator. This puts more passes within range of a boat in the arctic than one in the tropics. In any location though, the satellite data must be supplemented by other instruments between passes. A gyrocompass and doppler sonar are normally used together for this, but where doppler sonar is not effective, Loran-C or inertial navigation may take its place.

A gyrocompass, using the effect of the earth's rotation on a gyroscope to determine direction, provides the boat's heading. It doesn't depend on magnetism like an ordinary compass, so it isn't affected by magnetic declination or by metal on or near the boat.

Doppler sonar sends out pulses of sound to the sea bed in beams in four directions and receives them back, reflected from the sea bottom if the water isn't too deep. The doppler effect gives the sound received by the moving boat different frequencies from those sent out. The frequency is used to determine the boat's speed. And comparison of the four beams provides data on the direction the boat is headed. In water deeper than about 1000 feet, the sonar instrument receives sound, not from the sea bed, but scattered back from the mass of water. The scatter is from water that is moving in currents, so doppler sonar isn't very accurate in deep water. A velocimeter measures the velocity of sound in the water, to adjust the doppler sonar data. An inclinometer measures pitch, roll, and yaw of the boat for additional corrections to the doppler sonar.

In deep water, either Loran-C or inertial navigation can substitute for doppler sonar. Loran-C (Loran, LOng RAnge Navigation) is a system that uses broadcasts from fixed base stations on land. It covers a range of about 1500 miles, but does not yield locations accurate enough for seismic exploration. But for use between satellite passes its signals can be handled in a different way to give accurate data on change of location, rather than absolute location, of the boat.

Inertial navigation uses an inertial platform on which two pendulums are mounted at right angles to each other. Any change of speed of the boat in either a forward or a sideward direction causes the pendulum mounted in that direction to swing. The inertial platform on which the pendulums are mounted is held level by three fast-spinning gyroscopes mounted at right angles to each other. The gyroscopes' resistance to tilt in each of the three dimensions keeps the platform level, so the swings of the pendulums are not affected by the pitch and roll of the boat. Such a platform is interesting to watch. In staying level, it appears to tilt wildly as the boat rolls and pitches. Watching it for long could make a person seasick.

From all its data, the computer calculates locations of the boat between satellite passes. Then, at the next pass, it gets an update, a new satellite-derived location. The navigation data is recorded on tape. Later, in a computer center on land, more accurate locations can be calculated by smoothing the jumps in the data at updates and by other techniques, and a map made.

This type of satellite surveying works anywhere on Earth, but isn't as accurate as short-range shore-based methods. So it is used primarily for areas far at sea, where the shore-based methods won't reach. The only cost in using it is for the equipment and operators on the boat. It is ready for use at any time. Its convenience and economy make it a system to be used for small programs or programs that are to be shot quickly.

A newer system, NAVSTAR Global Positioning System, GPS, uses more than one satellite at a time, so comparisons of readings from different satellites give accurate locations quickly. The goal is to have 18 satellites in orbit, with at least two in range of any point on earth at any time. There are not yet enough of the GPS satellites in orbit to always have more than one in range, so the system now operates like TRANSIT in needing other data between satellite passes.

There is another use of either system of satellite surveying—to determine the positions of fixed points. Its application in offshore seismic exploration is in establishing base stations for other surveying methods. A portable satellite receiver is set up and left in one spot for several days to read a number of passes, maybe as many as 30. Then a best location is derived from all of them. In this application, satellite surveying is highly accurate. In addition to base stations, it is used to determine well locations and bench marks for surveying on land. Greater accuracy is obtained with the translocation method. In it, two receivers are used simultaneously, one at the location that is to be determined, the other at a known location. Any irregularities caused by changing atmospheric conditions that affect both can be corrected.

With Base Stations

There has been a proliferation of methods of doing radio surveying from base stations, which consist of radio equipment at known fixed positions.

Range-Range

In the range-range systems, at least two base stations are used to determine the location of a mobile station—the boat in offshore exploration. Each station has a transmitter and receiver. Each base station receives a signal from the mobile station, and immediately transmits it back. So equipment on the boat, by measuring the time from transmission of a signal to receiving it back, can determine the distance to a base station. From the distances to the two base stations, its own location can be determined. The distances can be represented on a map as circles drawn with centers at the base stations, with the boat located at the intersection of two circles. Actually, the circles intersect at two points, on opposite sides of a line connecting the base stations. But one of those positions is usually far from the expected position of the boat, and in offshore work it is often on land. It can be ignored.

There are several techniques for enhancing the accuracy of the survey. The base station locations are determined reliably and precisely by optical surveying, satellite, or other means. The equipment is calibrated by running the boat across a base line, the direct line between two base stations, and reading the distances to the two stations. Adding the two distances should give the already-known distance between stations. The equipment on the boat can then be adjusted so the distance reads correctly. The calibration is re-checked occasionally during the shooting by crossing the base line again.

If more than two base stations are used, then at any time a three-way fix can be obtained by reading the distances from three stations. This is a check on how well the locations from different pairs of stations agree. Three-way fixes can be obtained when needed, or at every position determined.

Like other triangulation, getting a location from two known points, range-range radio surveying works best when the two distances involved are about equal. Measurements made when the directions to the stations form an angle of less than 30 degrees with the base line are considered unreliable.

Instead of using the measured times to determine distance to two base stations, the differences between the times can be used. This requires at least three stations. The patterns of equal distances are hyperbolas instead of circles.

Phase Comparison

Phase-comparison survey systems use three or more base stations and a mobile station, like range-range does. But instead of using measurements of the time taken by the radio waves in going from one station to another and back, the phase-comparison systems depend on interference between two of the shore stations' broadcasts. The interference forms standing waves, fixed patterns of radio waves that can be detected by a receiver on the boat. The pattern repeats in cycles, and it is the different parts of a cycle that can be distinguished from each other. But the equipment can't tell the difference between comparable parts of different cycles. The strip of territory covered by a cycle is called a lane. This is another hyperbolic pattern. To avoid confusion between lanes, a lane count is kept, a count of how many lanes have been crossed since leaving some known point. However, any time the equipment fails, the lane count is lost. One way around that problem is to also broadcast an additional set of radio waves with a different interference pattern, with less accuracy but wider lanes to help identify the narrow lanes. Any other system used with the phase-comparison system can also identify lanes—satellite, Loran-C, whatever is available.

Forms of Base-Station Surveying

A great deal of ingenuity has gone into base-station systems. Shoran (SHOrt RAnge Navigation) originally used radio waves that traveled in a line of sight. But then extended range shoran, XR shoran, utilized radio waves that were scattered, so they could be received over the horizon. This kind is in use today. It is used mostly for the type of operation that calls for setting up base stations for a month or so, just for that program. To operate a base station, the equipment is moved in to the location, assembled, an antenna about 50 feet tall erected, a shelter set up if there is not one there already, and an operator left there to keep the equipment running 24 hours a day for a month or so. The operator hires a local assistant or two, to haul things and do other chores. The operator may not know the local language, and so may have a social life limited to radio communications with the other base stations and the boat, and maybe with other similar networks in the vicinity. It takes a special kind of person to be this type of base station operator and enjoy it.

One system lends itself to more permanent setups. A chain of stations is established, usually along a coast that has considerable activity. Antennas are taller, 100 to 500 feet. Range is about 80 miles. The system can be used by a number of boats at the same time. So the network is put up by a surveying company, which charges according to use. It can be used in several modes, range-range, hyperbolic, combination.

Another system uses radio waves that follow the surface of the earth. It too tends to be a system that may be set up fairly permanently. It uses antennas from 25 to 100 feet tall, and has a range of around 300 miles.

Acoustic Surveying

A way of surveying offshore for fairly small areas, especially in remote locations, uses acoustic transponders instead of radio. A transponder can receive commands by sound, and can send sound signals in reply. It is anchored and held a little way above the sea bed by some buoyant material.

Transponders are planted at several points in the area to be surveyed. Relative positions are determined by timing of sound signals from one of them to another. A satellite receiver on a boat can be used to obtain precise locations, by timing acoustic signals from the transponders while reading a number of satellite passes. Then a detail seismic survey of the area or a 3-D survey or a well location site survey can be made with positions of the boat determined by timing signals returned from the transponders.

This method can be used at any location. It does not need to be within range of base stations on shore. And its use of a number of passes eliminates the inaccuracy of normal TRANSIT satellite navigation for seismic work. There are some variables that must be checked—salinity and temperature of the sea water at the various depths from top to bottom of the sea.

Integrated Systems

Some survey systems are short range and very accurate, others longer range; some have one shortcoming, others another. To obtain maximum effect from the various systems, seismic boats often use integrated survey systems, that employ several different systems and use a computer to determine the best position from the combination of readings. Where one system has a problem, another may have good data. And where all are good, a location determined from all of them is better.

It is not always necessary to have a special receiver on board for each of the navigation systems. An integrated navigation system with a computer may be able to receive and use data from several of those systems with its own radio equipment.

Postplots

All of these systems for surveying at the same time as shooting, even integrated systems, have a limitation on the precision of locations that can be determined at the time. The systems can make comparisons and updates of data, but can't consider data that hasn't been obtained yet, and can't go back and rework the data when some new information indicates that it should all be changed.

These things can be done best after the survey is complete, in a processing center. This is called postplotting. Magnetic tape recordings of the survey data are sent to a survey processing center, just as the seismic tapes are sent to a seismic processor. The two may be in the same organization, or they may not. The different types of information are compared, wild jumps are smoothed out, updates are made to influence information obtained both before and after they occurred. This is similar to seismic data, which when shot is in minimum-phase form, but can later be re-formed into zero phase, with half of the energy of a reflection appearing at a reflection time before the precise reflection time.

The postplotting of survey data requires trying different corrections before determining the final ones to be used. It takes some time to accomplish all this. So for a large seismic program, it may be a month or so before a final map is available. This can be frustrating, as seismic data processors may have to wait for it before they can complete their work. But it is worth the wait to have reliable shot point locations.

Hydrophone

The type of geophone used offshore is usually called a hydrophone. This is something of a misnomer as it, like any geophone, obtains data about the earth (geo), not the water (hydro). And it is potentially confusing because there is another, more appropriate, use for the word hydrophone—a device to receive sounds in the water from fish, submarines, etc. But marine geophones are quite different from land geophones, so the word "hydrophone" became an accepted seismic term to make that distinction.

A seismic hydrophone is a pressure geophone. It detects, not changing acceleration as a land geophone does, but changing pressure. This is necessary to keep it from recording the motions of the water that an acceleration phone would record. Hydrophones use a piezo-electric material, that is, a substance that generates electricity when squeezed or released ("piezo", from a Greek word meaning to press). The compressions and rarefactions that are sound waves alternately squeeze and release, and the hydrophones generate electric current proportional to those pressure changes. Pressure phones work best at the pressure existing some distance down in the water. They are built into the seismic cable (Fig. 3-23), and towed at a controlled depth of around 40 feet.

The type of pressure phone used on offshore crews is the acceleration-cancelling hydrophone, which has two piezo-electric units in it, so placed that any effects of motion on one unit will be in the opposite direction on the other, and will tend to cancel. This fairly well eliminates the effects of boat engine noise, the cable being jerked, etc.

Marine Cable

A marine seismic cable, or streamer, is trailed behind the boat. The geophones—hydrophones—are built into it, so all the connecting of geophones and spacing within groups and from group to group is already done.

On long trips and when going into or leaving port, the cable is carried on a large reel on the back deck (Fig. 3-24). Then, at the start of a shooting program, it is payed out and is towed behind the boat throughout the shooting, day and night.

A conventional marine cable is thicker than a land cable, and is in a protective clear plastic flexible tube about 3 inches in diameter. In the tube can be seen the wires with their color-coded insulation. They are loose, not wound tightly together like in land cable. A steel cable in the center of the marine cable takes the strain of towing it through the water. Before a shooting project starts, the cable must be balanced, made to be of about neutral buoyancy. This is done on the boat by

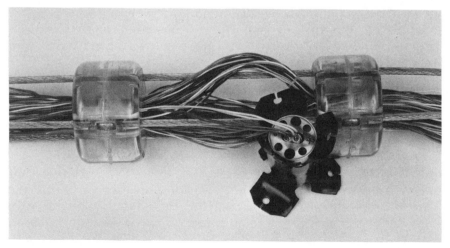

Fig. 3-23 Hydrophone in cable   *Credit: GeoSpace Corp.*

Fig. 3-24 Cable reel

filling the plastic tube with a kerosene-like liquid which is lighter than water (Fig. 3-25), and by taping flat strips of lead onto the outside of the tube where needed. The balancing can take most of a day.

The cable is in sections, so a faulty one can be taken out and replaced. Some sections have geophones and others do not. These dead sections are placed between sections containing geophones to achieve the desired spacing between groups.

A newer type of cable has electronic units along its length to arrange the geophones into groups as needed at the time, rather than having them already in fixed groups.

Also some cables have built-in distributive units that transmit data in digital form. The cables, like similar land cables, need only two main conductors to carry the data sequentially. The two conductors can be wires, or glass fibers that convey the signal as light, just as with distributive systems used on land.

Depth controllers are put on the cable at intervals, about four of them on a cable. A depth controller looks something like a little torpedo, about two feet long, with two short winglike fins. It separates lengthwise into two halves, hinged together so it can be opened and then clamped shut around the cable. The fins can be angled like the diving planes of a submarine, to keep the cable at the desired depth for shooting.

Fig. 3-25 Adding fluid to cable for buoyancy

There may be magnetic compasses at a few places in the cable, to indicate the angles at which the parts of the cable trail under the influence of water currents. This information, combined with the known lengths of cable to those units, is used to determine where the geophone groups are in relation to the boat. This information can be useful in 2-D shooting, and is essential in 3-D work.

There are elastic stretch sections ahead of and behind the main cable to reduce the jerking from sudden motions of the boat or the tail buoy at the end. The tail buoy is attached by a length of rope. It is usually a catamaran raft with plastic foam floats. Its purpose is to be detected from the boat, for determining the direction to the end of the cable. So it is equipped to be detected with bright color, radar reflector, radio responder, or radio transmitter.

The tail buoy is usually from a mile and a half to over two miles behind the boat. While shooting, the cable itself is not visible, as it is down in the water. By looking hard, the tail buoy may be barely seen as a sparkle on the horizon.

When changing from one line to another, the boat runs in a long curve to get the cable aligned straight behind it for the new line. A big loop is run if, in the course of shooting a line, something causes the shooting to stop. Then, when the cable is again straight, some of the line is re-shot to overlap the earlier part, and the line continued. The boat runs day and night, different shifts of people taking over, the pop or bang of the energy source every few seconds interrupted only for line changes, or sometimes for repairs that can't be made while shooting or during line changes.

To record data under an obstacle like a platform or a small island, the cable may not be used, but instead radio telemetry units on anchored buoys connected to strings of geophones on the bottom may be substituted for it. The telemetry units are placed in a line on one side of the obstacle. The boat runs on the other side, firing its source of energy, receives data from the geophones by radio, and records information from midway between the geophones and source.

Green Dragon

There is a temptation to paint teeth or other ornamentation on the tail buoy, especially as, if all the electronics fails, bright colors may help to spot it. It is a mile or more behind the boat, and there is no visible connection between the two. When the boat is making a loop, the tail buoy might even be seen going by in the opposite direction. It appears to go where it will, somehow under its own power, independent of winds and currents. One offshore crew reported that an official advice to mariners contained a warning about a green dragon that had been sighted in the sea.

Marine Sources

There are a number of ways to produce seismic energy offshore, and a special problem to contend with. The problem is the bubble effect—the repeated expansions and contractions as explosion and water pressure alternately dominate. The bubbles from a marine source must be measured so they can be countered in data processing, or the source designed to have little or no bubble effect. This concern with bubbles runs all through marine source design.

Marine Air Gun

The seismic source most used offshore is the air gun. It is a chamber that is filled with compressed air, and the air released suddenly when needed. The pop of the released air is the seismic impulse. The air does produce repeated bubbles. Some air guns use a wave-shaping kit, a second chamber into which some of the air goes, to be released just enough later to be at the same time as a contraction of the first bubble. It doesn't eliminate the bubbles, but reduces their effect.

A number of air guns are trailed in an array, with hoses extending from them to air compressors on deck. There are about four large compressors and maybe a dozen air guns. Several arrays may be used, like pattern shooting on land (Fig. 3-26). Air is pumped into the guns. Then the air in all the guns is released at the same time or nearly the same time. The release is a pop. The pops occur at about ten-second intervals while the seismic line is being shot, which may take several hours. Then, during the run to the start of the next line, if necessary, the guns are hauled on deck and serviced.

The guns are trailed a little way behind the boat or beside it, suspended at a depth of around 30 feet below floats. They may be attached one behind the other, or may be trailed separately, or even attached to a large framework that is trailed.

The guns are usually arranged in a tuned air gun array. That is, the guns are of varying sizes, spaced varying distances apart, and have varying time delays—all to cause the expanding air from the different guns to blend, making in effect one large, sort of flat, detonation. The different sizes of air guns produce bubbles that expand and contract at different rates, so the combination of them all allows the bubble effects

Fig. 3-26 Several arrays   *Credit: Geophysical Service Inc.*

to partially cancel each other. But even so, it is useful to record the bubble pattern with a geophone near the array every time it is shot, to help the processors reduce the bubble effect on the processed data.

The effect of air gun firing that is noticeable from the boat is a rather sharp shock, somewhat as though someone had hit the hull with a padded hammer. A few seconds after the pop, the water sort of boils up in a small area, and looks white from the addition to it of many tiny air bubbles. Then 9 or 10 seconds after the pop, the next pop occurs.

Gas Gun

A gas gun fires a mixture of propane and oxygen, or propane and oxygen-enriched air, in a chamber. In offshore operations, an array of gas guns is trailed from the boat, much as air guns are handled. A tank of propane is carried on deck. A tank of oxygen may also be carried, or a nitrogen-removal process may be used instead to increase the percentage of oxygen in air drawn from outside.

In one gas gun system, the explosion takes place in a heavy rubber sleeve that expands with the explosion, then contracts, letting the gases escape through a tube to the air. This reduces the expand-contract-expand bubble effect.

Other Sources

Several other seismic sources are used offshore, but to a lesser extent than air guns or gas guns.

Steam from a boiler on deck is suddenly released from a chamber in the water. The particular advantage of this source is that the released steam quickly condenses into liquid water, so there is no volume of gas remaining to form later expansions of a bubble.

Water guns, another source designed to eliminate the bubble effect, expel water from a tube into the sea at high pressure. The fast-moving water pushes seawater aside, leaving a partial vacuum behind it. Water rushes in to fill the space, producing sound. This is an implosively created sound—like a thunderclap, which results from air replacing the air that was pushed aside by lightning.

Another method uses very small charges of an explosive. They are dropped through a tube into the water, about as deep as other sources. Each charge detonates right after leaving the tube. This method has more concentrated energy than the other sources, so it is used where deep information is difficult to obtain.

Occasionally, in special circumstances (like some shallow water operations) larger charges of explosives are suspended from expendable floats a few feet below the surface of the water and detonated. A geyser of water shoots up as the gases blow out of the water before a second bubble expansion can take place. This was the original type of source for offshore seismic exploration, and was the only one used for many years. Its use is now forbidden in many areas.

Recording

Offshore recording uses the same type of magnetic recording equipment as is used on land, but it is designed to fit the requirements of shooting in rapid sequence, on a 24 hour basis, and at sea.

There is more equipment. There are usually two or more tape drives, so a switch to a new reel of tape can be made without halting the operation while reels are changed. There is a larger stock of tape. There are indicators for cable depth, condition and depth of air guns or other source, etc. There are enough personnel to work on a shift basis. There are enough spare parts, and enough room, to make repairs while shooting is going on, or on the run between lines.

The doghouse is fairly large, although it may have quite a lot of equipment and people. It is air conditioned, not primarily for the comfort of the people, but so the heat-producing and heat-sensitive electronic equipment is kept at a good working temperature.

In normal operations, several people are in the doghouse, watching instruments, taking notes, maybe repairing things or improvising to make things handier. There is a dull boom, not loud, of the air guns or other source being fired. The tape reels advance a few inches and an oscilloscope shows the motion of the traces as they receive reflections. In less than 6 seconds, usually, all the instrumental activity of recording that shot is over. People may still be writing about it on logs of the operation. Then, about ten seconds after the boom, there is another boom. Another shot has gone off. The tape advances.

The written logs give such details as depth of the guns, depth of the cable, time of day, sea state, shot number. Some recording equipment also puts much of that information and position data on the tape, so it will be readily available to data processors. Every few shots, a monitor record is made and inspected to see how well the recording is doing.

A plotter may be making a seismic section as the shooting proceeds. It plots one trace from each shot. The section slowly lengthens as the boat moves along the line, showing the subsurface as it is being recorded.

After a number of shots are recorded, the takeup reel is full. Someone switches to another tape drive that is already loaded with a full reel and an empty takeup reel. The full tape is taken off its machine, labeled, put in a round plastic case, the case labeled and sealed and put in a plastic sack, the sack sealed and placed in a carton that tape came in, notes made about that reel of tape. Someone puts another full reel of tape on the drive, threads the end through a number of tension rollers and onto the empty reel, which is put in the takeup position.

Meanwhile, more shots, on and on for maybe hours, to the end of that line. Then the end of the line with its sudden absence of the sound of shots. The boat swings in a big loop to get the cable straight behind it for the start of the next line.

The recording people are also the ones who look after the cable, laying it out behind the boat (Fig. 3-27), balancing, testing. And, when an instrument in the doghouse indicates that the cable has a break or similar problem, they come boiling out onto the back deck, reel it in to the bad spot, see what the trouble is, repair it or replace a segment, run the cable back out. They communicate with the wheelhouse and recording room by intercom, and on some boats the helmsman and anyone back in the recording room can watch them on closed-circuit television.

Sometimes several segments of cable are badly damaged, say from

Fig. 3-27 Reeling cable   *Credit: Geophysical Service Inc.*

being dragged across a coral reef and acquiring many punctures. Then a cable party may be necessary. Anyone who is willing is welcome at the party. That may include recording people who are off duty, people in other departments, whoever strongly wants to keep the work going smoothly or finds it more fun to pitch in than not. The bad segments are removed and replaced, if there are enough spares on board, so shooting can continue. The damaged cable is piled or coiled on some open space, often the helicopter landing area, and the leaks repaired. The only way to get it up there and to move it from one coil to another as repairs are made is laboriously by hand. The cable is heavy. It is oily, from the kerosene-like flotation fluid that leaks out. The fluid can burn the skin, and there isn't any shelter from the sun up there. It makes the return to normal operations feel like a vacation.

In the recording room when things are running smoothly, at shift-change time the relief crew, having just eaten, wander in carrying cups of coffee and pieces of pie. They report on what's for dinner, find out what has happened and how things are going, and take their places. The shift going off stand around a while, glad to be off duty, but some-times a little reluctant to leave the one place on board where the action is.

The Signal and Its Signature

The signal is the sound that starts its trip into the earth, not the response that comes back to the surface. The signal is made up of whatever sound the source produces. The shape of a recording of the signal is the signature of that source. This is not the same as a wavelet, which is the same signal after it has been distorted by a trip down into the earth and back up. A seismic section is made up of wavelets.

The main importance of seismic signatures is in offshore work, where the sound is affected by the repeated expansions of the bubble produced by the explosion. The signature then, shows the initial impulse and the bubble expansions following it, with their relative strengths. The goal in source design is to get a strong initial impulse with weak, or no, bubble effect.

A signature offshore can be recorded in special tests by placing a geophone where it can record what is going into the earth before it reaches the ground under the sea. This is done in two parts, with a geophone a few feet away from the source to give a near-field mea-surement, and by a geophone farther away (over a boat's length away) giving a far-field measurement. Deep water is required for these tests, so the reflection from the sea floor will not appear on the record until the complete signature is recorded (Fig. 3-28).

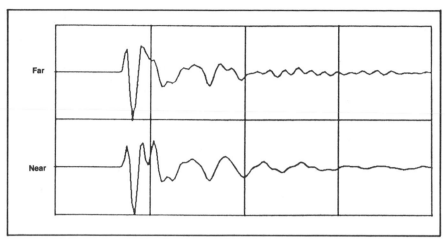

Fig. 3-28 Far- and near-field signatures   Credit: Digicon Geophysical Corp.

A normal signature is one strong sharp peak and trough, followed by several others of diminishing strength until they get too small to be noticed. This is what is put into the earth, so any reflecting horizon bounces the whole thing back, distorted into a wavelet. A seismic section that is made up of these long duration reflections has a lot of repetition in it, and overlapping reflections. The more the first energy of each reflection can be made to dominate, and the more the following parts can be subdued, the clearer the section will be. So it is desirable to design sources to have a strong first impulse and relatively weak bubbles (Fig. 3-29) or no bubbles.

Other Surveys

In addition to normal 2-D seismic reflection lines on land and at sea, there are a variety of other kinds of investigations. Three dimensional seismic surveys are becoming more common, and may one day displace 2-D as the usual type of seismic work. An LVL crew gathers refraction information for a surface-source crew. Shallow-water shooting ties offshore and land shooting together. Engineering surveys are necessary before offshore wells can be drilled. Gravity and magnetometer are non-seismic geophysical methods. Shear wave and refraction exploration are discussed in Chapter 10, Other Seismic Methods.

3-D Shooting

When seismic data is obtained in separate straight lines, the information is essentially two dimensional, more or less in a plane below the

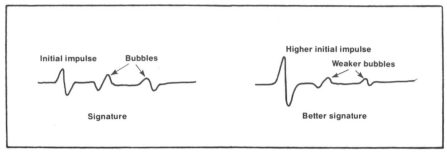

Fig. 3-29 Marine signatures

line. But the subsurface is three dimensional. So 2-D shooting presents problems when there are reflections from the side (is a certain reflection from below the line or from one side?), when an anomaly is found on a line (which side of the line is the main part of the feature?), when faults are found on the section (which map direction does the fault trace go?).

There are several ways to obtain three dimensional data in the field. All take advantage of the fact that subsurface data is obtained from midpoints between source and receiver, by arranging shots and geophones so that the midpoints are not just in the usual straight lines, but spread out over some width to the sides of the line or giving blanket coverage of a wider area. Although the principle is the same, the methods used on land are necessarily different from those used offshore. Land 3-D will be discussed first.

A partial form of 3-D shooting can be obtained by laying out two or more lines of geophones parallel to each other with a line of shots parallel to them. The reflection points will be in two lines between the lines of geophones and line of shots (Fig. 3-30). Or there can be two or more lines of shots and one of geophones, with the same effect. This gives some width to the data. In another method, a line can be shot with geophones and shots in a crooked path, perhaps following a trail or river. A midpoint between a shot and a geophone that is around a bend from the shot will not be on the line. There will be a scattering of points near the trail (Fig. 3-31). So again the data is spread over some width.

A true 3-D survey covers an area. This too can be accomplished in different ways. A geophone cable can be laid out completely around an area, and shots fired along the cable, so they too go around that area. The midpoints will mostly be within the area, rather than on the line. The area can be a rectangle or an irregular shape (Fig. 3-32). The whole shape is filled in with data. This method provides enough data for CDP

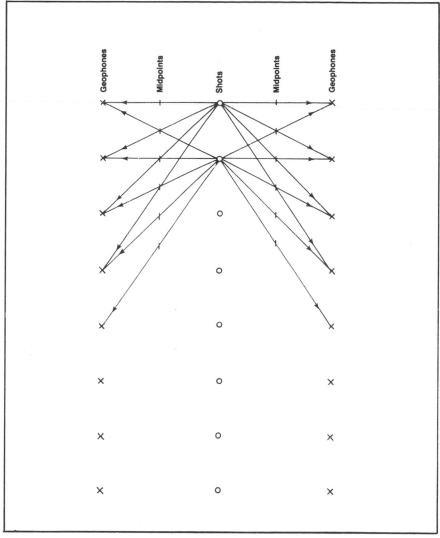

Fig. 3-30 Double line

over the area. But the depth points do not coincide. So binning is resorted to. Small rectangular areas are established, each one a bin for CDP stacking. All traces whose depth points fall within a bin will be combined to make one stacked trace.

There is a disadvantage of this kind of 3-D shooting. Some interior parts of the area have data only from long traces (traces from receivers far from the source). The depth points are not close to either shot or

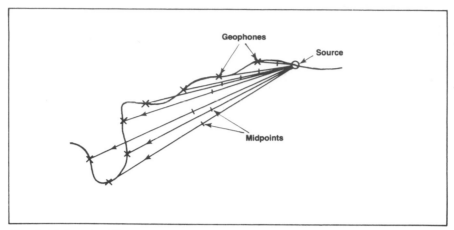

Fig. 3-31 Crooked line

receiver. Better results can be achieved if each common depth point has a mixture of long and short traces, as in normal CDP shooting of 2-D lines.

When a shot is fired to one side of a line of geophones, data is obtained from a line of points parallel to the line of geophones, half as long as that line, and midway between it and the shot (Fig. 3-33). At present the most common way of shooting 3-D uses this principle, with several parallel lines of geophones and one line of shots perpendicular to the geophone lines. That gives 3-D CDP coverage of a rectangular area (Figs. 3-34 and 3-35). Then the same process can be repeated for an adjoining rectangle. An area is shot in these rectangular units. The lines are typically close together—about two to four times the distance from one geophone group to the next on the same line. They make a blanket coverage of the area without losing the advantages of 2-D CDP shooting. Similarly, 3-D data can be obtained with the lines of shots and geophones all parallel.

A grid is carefully surveyed and staked (Fig. 3-36). The surveying must be done as an area rather than as isolated lines. The seismic operation is larger and more involved than separate 2-D lines. Any failure of one part of the operation can delay all the other parts, adding to the expense of the survey. So it must be very thoroughly planned and closely supervised. Timing of the different parts of the work must be coordinated, so one part of the crew doesn't stand by waiting for another part of the crew to finish.

Offshore, the most common way of shooting 3-D is to use closely spaced parallel lines. But the unavoidable feathering of the cable, that

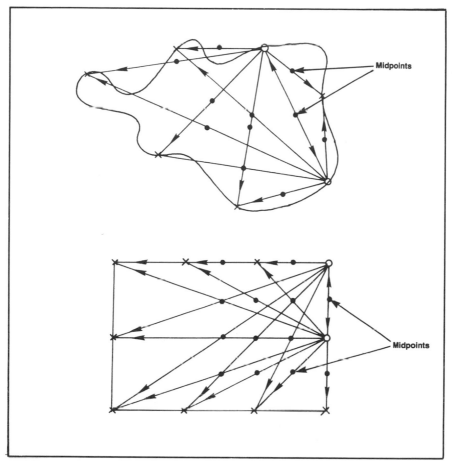

Fig. 3-32 Enclosed areas

is, the bend of the trailing end caused by currents in the sea, puts the depth points not exactly on the lines that the boat follows. The feathering causes the midpoints for the far geophones to be offset to the side of the cable, while the midpoints for the near geophones are closer to the cable. So as the boat advances, there is, not a line, but a band, of midpoints (Fig. 3-37). The exact offset of the cable can be determined by magnetic compasses installed in the cable, that transmit information to instruments on the boat. With length of cable and direction known at several points, geophone positions can be determined, and midpoints can be calculated from them. The known midpoints are then put in bins as on land. If the feathering is great enough, midpoints

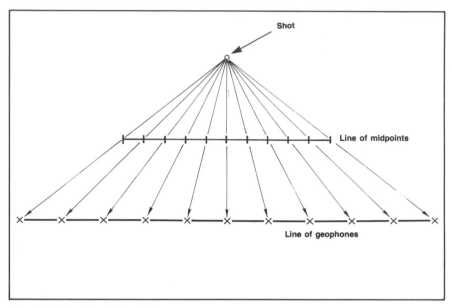

Fig. 3-33 Shot to side of line

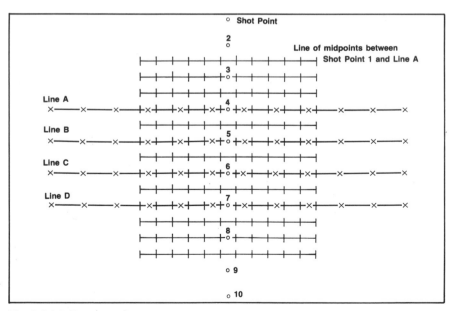

Fig. 3-34 3-D rectangle

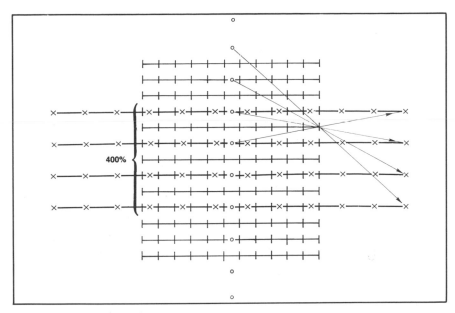

Fig. 3-35 Common midpoint

from several lines will fall in the same bins. This gives a useful mix of near and far traces in a bin.

Alternatively, instead of just using the feathering of the cable with its position determined by compasses, the swath of subsurface points can be made even wider by shooting two sets of air guns that are some distance apart. The two sets of guns are equipped with paravanes, devices like rudders, that push them out to the two sides of the boat's path. The two sets are fired alternately. There are no more shots than normal, but the depth points are spread out over a wider path. This gives wider 3-D coverage but with a smaller degree of stack.

When there is an obstacle to shooting, like a production platform, the 3-D coverage of the area will not be complete when the shooting is done the normal way. The gap can be filled in with a two-boat operation. In its simplest form, one boat has the cable and recording equipment, and the other boat has the airguns or other source. But both must have radio surveying equipment. The two boats go on parallel courses, on opposite sides of the obstacle. One boat shoots and the other records. The midpoints, between them, fill in the gap, even directly under the obstacle. There are variants of this method. Both boats may fire sources alternately, yielding two lines of data, but of

*Fig. 3-36 Laying out 3-D—white lines added to picture to show
planned grid   Credit: PRAKLA-SEISMOS GMBH*

lower CDP. Both may have cables and sources, so both record the
alternating firings. All two-boat operations require good coordination.

When the common depth point process was developed, it was radically different from conventional, 100%, shooting. CDP called for taking many more shots per mile, use of more cable and more recording
capability. It was a bigger operation that required a higher level of organization of the field crew to keep everything running smoothly so as to
avoid expensive standby for parts of the crew. It also called for a more
intensive effort in processing. All around, it was a more expensive
method. But the advantages of CDP data were so great that in a few
years, almost all shooting was CDP. The industry is so geared to CDP
that 100% shooting now would probably not be any cheaper.

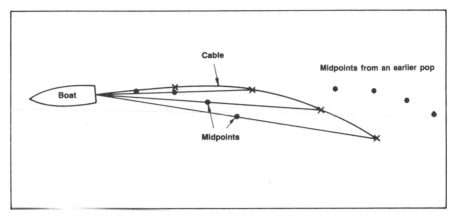

Fig. 3-37 Feathered cable

Three-D shooting is at present in the stage that CDP was in shortly after it was developed. Three-D is so expensive that it is primarily used for field development, and very little for exploration. It is a bigger operation, requiring a larger crew. But it yields far better data than 2-D. So it is likely that in a few years 3-D will be the conventional type of seismic exploration. Although it now seems impossible, it may also soon be that shortcuts and efficient ways of shooting will make 3-D as cheap as 2-D.

When 3-D becomes the most common type of seismic exploration, it will have solved many of the major problems we normally concern ourselves with. But in so doing, it will not reduce the number of problems in seismic interpretation. Relieving us from problems like sideswipe and fault direction, it will allow us to be involved more with things like seismic attributes and resolution.

Shallow Water, Surf, Mud Flats

The transition between land and deep water, the two main shooting areas, is a problem zone. Subsurface features aren't neatly divided by the present shoreline. They may extend from under the land to under the sea. Special field techniques are needed to cover this zone, and to link the land and water shooting together (Fig. 3-38). It has its own problems, of both difficult access and seismic noise.

There are special shallow-water crews, on shallow-draft boats. The surveying used may be the usual radio surveying of offshore work, or it may be a more accurate short-range radio system, using nearby base

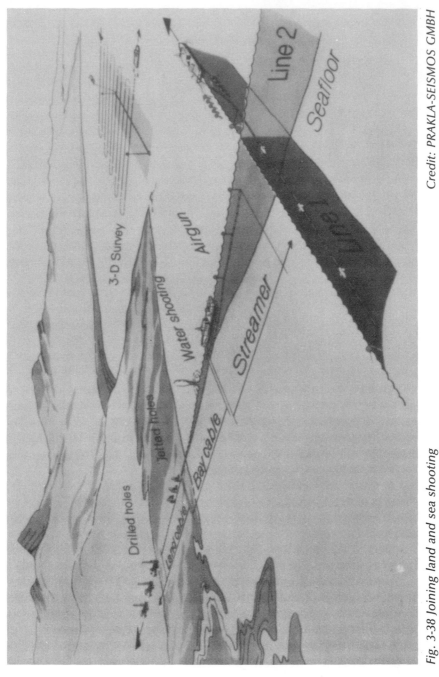

Fig. 3-38 Joining land and sea shooting

stations. In some cases the surveying may be by transit or theodolite and electronic range-finding from stations on shore.

On mud flats or in water just a few feet deep, an essential land-type program may be carried out, with people walking or wading depending on the tide, drilling holes, placing charges in the holes, laying out cables, and planting geophones. The holes may be jetted by people-portable pumper drills, or by larger rotary units mounted on shallow-draft barges. Or, instead of explosives in holes, detonating cord may be laid out. When fired, it produces a long, but otherwise small, momentary fountain. Its length also serves, like in-line groups of shots or geophones, to cancel some horizontally traveling noise.

Radio telemetry can be installed on buoys that are placed in the water. The geophones of the group are connected to the transmitter by wire. With this method data can be obtained from shallow water up onto the beach, to tie lines shot on land.

The surf is a source of noise, crashing over the geophones that are planted in its way. However, much of this noise is outside the seismic frequency range, and can be pretty well eliminated by filtering out the low frequencies.

Engineering Surveys

Offshore, when a well is to be drilled, there are a number of conditions at and under the sea floor that could be hazardous to the rig. So a special kind of investigation is conducted before the rig arrives at the location. This is a wellsite survey, also called a hazard survey or just a site survey. It is a type of engineering survey. Engineering surveys include all sorts of shallow investigations for construction, dredging, laying pipelines. They include several kinds of information about the sea bed and the shallow subsurface. They are also sometimes run as an aid to interpreting a standard offshore program.

A wellsite survey is required by the rig owner's insurance company before the rig may be moved to a drilling location. Although it is a geophysical survey, it is usually contracted for and overseen by drilling personnel. But sometimes a geophysicist may be called upon to supervise it. It usually consists of a survey of the seabed over a square that has the proposed location at its center. Normal requirements are for fairly closely spaced lines of readings of several instruments: water depth recorder, magnetometer, side-scan sonar, spark- or other-type device for obtaining shallow seismic data.

A depth recorder uses reflected sound waves to show the depth of the water under the boat. Where the sea bed is soft, the recording may show the bottom and also a harder layer underneath.

Side-scan sonar is a sound-reflection system that gives three-dimensional data. A "fish" looking something like a model submarine is towed behind the boat. Sound from a pinger on it fans out, reflecting from a swath of the sea bed. A detector on the fish receives signals from the swath on both sides of it. These are recorded on a continuous strip showing, usually, all the sea bed from directly under the boat to 500 feet to each side (Fig. 3-39). The recording shows rocks, ripple marks in sand, sunken vessels. It looks something like an air photo over land, but it is a negative image and is "illuminated" from the center line. The image is in a reddish-brown color. Objects that stand out are dark. Shadows are white, and extend outward from the objects casting them. Height of an object can be estimated from the length of its shadow. Rulings are made on the record parallel to the boat's heading, to indicate distance to the side of the boat. They are normally in 50 foot increments. But the record can, at a different setting, be made to cover only 250 feet of sea bed on each side, showing more detail; or 1000 feet, to cover more area with less detail. In each case the lines are placed so they represent 50-foot widths of sea bed. Some equipment is metric, with 25-meter increments and a 500-meter width.

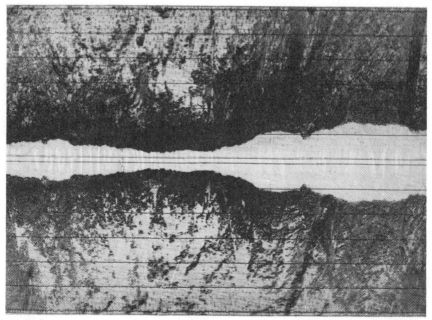

3-39 Side-scan sonar record   *Credit: E. G. & G. Environmental Consultants*

Side-scan sonar tells much about the bottom: hard or soft, mud, sand, rock, etc. If run simultaneously with a seismic survey and if the bottom has recognizable features, it can be a check of surveying at line intersections. It introduces to offshore work the landmarks that are so taken for granted on land.

There is another advantage of side-scan sonar run while shooting is going on. It shows the configuration of the bottom along the seismic lines. Each time there is a pop of the seismic energy source, it is recorded on the side-scan sonar record as a strong line across the recording. This interferes a bit with the picture of the sea bed, but the interference is more than made up for by the definite tie between points on the sonar record and the seismic pops. Every so often, someone identifies a pop by marking its number on the sonar paper record. Then, by counting from the identified ones, the location of any pop can be found, clearly marked on a strip map of the bottom. If the sea bed is not just mud and sand, but has some recognizable features, the sonar tape can be a great help in identifying locations, checking intersections of lines, positioning wells, etc.

Side-scan sonar can clearly show the bubbles of a gas seep. This helps in locating near-surface hydrocarbons, to warn against the blow-out hazard of drilling into shallow gas.

The electrical spark method is a seismic exploration system that generates sound by firing a spark, and has a short seismic cable. It uses sound of a higher frequency than normal seismic energy to yield very detailed information for a short distance down under the sea bed. In general, the less penetration the more detail. The very shallow seismic surveys can show remarkable detail on small faults that extend up to the sea bed. This is another technique that is useful in areas where shallow gas pockets can be a hazard to drilling.

All the shallow seismic reflection methods use short geophone cables—necessarily, as the geometry of shallow data does not allow for long offsets. That is, distant receivers will not receive reflected energy from shallow horizons, but only refracted energy. The short cables are often referred to as mini-cables. Such a near-surface survey may be run at the same time as a normal seismic survey. The mini-cable may be, for its entire length, nearer to the boat than the nearest geophone group of the regular cable. The mini-cable and its associated equipment record data from the near-surface in fine detail. The normal seismic cable obtains the usual data, with maybe faults, or structure, approaching the surface, but not clearly shown at the top of the section. The mini-cable data then can provide interesting answers to geological questions. A fault may be seen at the surface, and be deduced to be the same fault as one on the normal section—projecting downward on one section and upward on the other. Or, equally useful, the deeper fault may be seen to not reach the surface. Similarly, structural highs and lows may or may not be found to reach the surface.

Gravity and Magnetometer

Gravity and magnetic measurements are not usually, on land, made along with seismic exploration. They are generally considered preliminary, to isolate prospects for seismic investigation. But offshore, to make use of the boat and the surveying and other information that are being provided anyway, one or both may be an integral part of a seismic operation. In this application, they are supplemental to the seismic work, and can assist its interpretation. Different approaches, not having the same defects, can serve as a check on each other.

A gravity meter, a gravimeter, uses a spring to weigh a small heavy weight that is built into the instrument. If dense rocks are nearer the surface than normal, the object in the instrument will weigh a bit more there than it does when weighed at another location. So the gravity meter indicates places where denser, usually older, rocks are near the surface, therefore some indication of structure. A structural high appears as a high gravity reading, a gravity high. The low density of a salt dome shows up as a less than normal gravity value, a gravity minimum. On a boat, a gravity meter is placed on a surface that is level regardless of the motion of the vessel, like the level platform of inertial navigation. The rise and fall of the boat and other effects of its travel are measured by various instruments, to later be taken account of in computer processing of the gravity data.

A magnetometer is, in effect but not in construction, a magnetic compass on its side. That is, it measures variations in downward magnetic attraction at different locations. Igneous rocks are more strongly magnetic than sedimentary rocks, so a magnetometer gives information on depth to basement, the bottom of the sedimentary rocks. On a boat, the magnetometer is trailed in the water.

CHAPTER **4**

Recording

The recording operations of seismic crews on land and offshore have been discussed as activities in the field. The way the data is actually put on tape is the subject of this chapter (Fig. 4-1).

Analog and Digital

Seismic data, or any other kind of data, can be recorded and processed by analog methods, or digitally. Analog refers to representing an amount of something by an analogous amount of something else. A thermometer, a clock hand, a seismic trace on a wiggle-trace section, are all analog. Temperature is represented by the height of the liquid column, time by the position of the hand, movement of the ground by the swings of the trace. Analog information is continuous, so between any two positions of the indicator, there are other positions that can be read. Having a quantity in digital form just means that it is given as numbers, digits, as discontinuous units, with no finer readings given between two successive numbers. A weather report gives temperature digitally. A calendar is digital. A seismic trace is represented digitally by a long series of numbers.

A line graph is an analog representation of data. A list of numbers is a digital representation of data. If you read the heights of the chart at intervals and write those numbers in a list, you have converted the data from analog form to digital. But if you read the numbers in a list, plot them on graph paper, and connect the points with a smooth curve to make a graph, you have converted the data from digital to analog.

In the early days of seismic exploration, recording was completely analog. The ground shook, a voltage was generated by a geophone, and a trace wiggled. When magnetic tape recording was developed, it enabled a trace to be replayed as though the shot had occurred again, and the trace could be re-recorded onto a seismic section. Still analog. The trace could be modified by different electrical processes, filtering out unwanted frequencies, changing amplitudes, shifting or stretching the trace.

109

Fig. 4-1 Recording instruments
Credit: Petty-Ray Geophysical Division, Geosource Inc.

Digital seismic recording at first seems like a backward step, in that it leaves gaps in the otherwise-continuous data. At intervals, the amplitude of the trace is recorded as a number. The parts of the trace between these samples are lost. However, the sample interval can be chosen as close as necessary. Sample rates of 4, 2, 1, and 1/2 ms (millisecond) are used. So it is safe enough to assume that the trace follows a smooth curve between samples. In exchange for that one loss, there are much greater advantages provided by the ways numerical data can be handled. Digital recording has a greater dynamic range, that is, difference between the largest and smallest amplitudes it can record, than analog recording. And with the data in the form of numbers, modifications can be performed by arithmetic. A trace can be replaced with another trace that is the result of arithmetic operations performed on the numbers of the original trace. For instance, to get a trace with twice the amplitude of the original trace, just multiply each of the numbers that make up the original by two. The things that can be done arithmetically require only that a formula be worked out and put in the form of a computer program, so a computer can do all the tedious arithmetic, its high speed making the process feasible.

After the originally analog data is recorded and processed digitally, it is converted back to analog form as traces. Then it can be a seismic section, ready to be interpreted, instead of just a list of numbers.

Aliasing is a characteristic of digital recording, as it is of any method of sampling data at intervals. It is the false picture you get if you don't sample often enough. The finer the detail in the data, the oftener you need to sample to avoid aliasing. If seismic data is sampled less that twice per cycle (peak and trough combination), that is, if there isn't at least one sample on each peak and one on each trough, the peaks and troughs that were not sampled will be left out. Then, when the wave is later reconstituted from the samples, it will form a smoother curve with a lower frequency (Fig. 4-2).

So, for a certain sample rate, say 4 milliseconds, some high frequencies (higher than seismic information) will be aliased, making them appear to be some of the lower frequencies that are being used for interpretation. To prevent that problem, an alias filter, or anti-alias filter, is applied before the data is sampled and recorded, to eliminate those high frequencies.

Binary numbers

Digital recording of seismic data uses the binary number system, that is, a system of numbers to the base two. Our usual arabic numerals (used in the same form by nearly everybody in the world except the

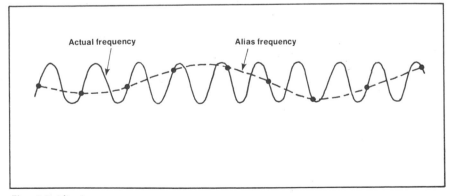

Fig. 4-2 Aliasing

Arabs—but derived from theirs) are to the base ten. We have a different digit for every number smaller than ten, and make up all larger numbers by combinations of those digits. So we make the number ten with the digits for one and zero, in that order. Then on through 11, 12, etc. Presumably the counting system was based on ten because we have ten fingers (fingers too are called digits). If there was a tribe somewhere that didn't use their thumbs for counting, but used a similar system, their numbers might be 0, 1, 2, 3, 4, 5, 6, 7, 10—the one and zero representing the number of fingers (not counting thumbs) on two hands, the number we call eight. Two and zero would be their way of writing sixteen, 25 would be twenty-one, etc.

Following this idea, a number system can be based on any number. All you need is enough different digits for all the numbers up to the base, then use 10 for the base, 100 for the square of the base, etc.

Numbers based on ten are called decimal numbers (from Latin decem, meaning ten). Other types are useful. Duodecimal, base twelve, can be divided up into more whole-number parts, so we buy eggs by the dozen. We can get a half, third, quarter, or sixth of a dozen without breaking an egg. Octal, the base eight already described, is handy for some computer uses.

Binary (we can't call it dual, that word's already been taken for another meaning), base two, is an especially useful system. Like the other systems, it has digits for all the numbers below the base. That is, it has the digits 0 and 1. So the numbers go like this:

Binary	Decimal
0	0
1	1
10	2
11	3
100	4
101	5
110	6
111	7
1000	8
1001	9
1010	10
1011	11

Now this binary system uses a lot more digits to write a large number than the decimal system does, but it has the big advantage that it needs only two different digits, 0 and 1. This makes it admirably suited to computer operations. The 1 can be represented on magnetic tape by a magnetized spot, the 0 by a spot that is not magnetized. Then with the simple convention of dividing the area of a tape into specific spots, and either magnetizing them or not, any number can be "written" on the tape. This is much more definite than a system like analog recording, that uses degree of magnetization.

Some other number systems, also using the binary on-or-off principle, are used in computers. One such is binary-coded decimal, which writes in decimal form, but by expressing each digit of the decimal number in binary form. As the tabulation above shows, four binary digits are enough to write any one of the single decimal digits 0 through 9.

But those four binary digits are enough to represent more digits than that. They can represent as many as sixteen separate digits. A number system with sixteen digits is called hexadecimal. It is more economical of space on magnetic tape to use those four to represent hexadecimal numbers. For the convenience of people, not computers, hexadecimal is written with sixteen different digits: 0, 1, 2, 3, 4, 5, 6, 7, 8, 9, A, B, C, D, E, F. Not having already-designed digits above 9, letters are used. We already know their order. So there are these equivalents between decimal and hexadecimal:

Decimal	Hexadecimal
0	0
9	9
10	A
15	F
16	10
18	12
31	1F
32	20

Exponential Notation

In dealing with very large or very small numbers, a lot of zeroes get in the act. The number 100,000,000 or .00000001 has so many zeroes that you may have to count them to decide what the number is. There is a different system of notation that includes a number that indicates how many zeroes there are, like "1 followed by eight zeroes", or how many decimals, like "1 in the eighth decimal place". Also, the system is so designed that it takes care of more complicated numbers like 132,000,000. It is the exponential system, in which that last number is 1.32×10^8.

The exponent, the 8 in that example, is a power of the number ten, the number of times that ten is to be multiplied by itself. Ten squared,

10^2, is 10 × 10, or 100. So a million is 10^6, and a hundred million is 10^8.

Negative exponents represent dividing by 10. So 10^{-1} is 1/10 or .1. And 10^{-2} is 1/(10 × 10) or .01. And 10^{-8} is 1/100,000,000 or .00000001 or "1 in the eighth decimal place". The negative exponent is the number of zeroes in the denominator of the fraction (e.g., 10^{-3} = 1/1000) or one more than the number of zeroes in the number when expressed as a decimal (e.g., 10^{-3} = .001).

A tabulation will help show the relationship.

1×10^3	1000	4.2×10^3	4200
1×10^2	100	4.2×10^2	420
1×10^1	10	4.2×10^1	42
1×10^0	1	4.2×10^0	4.2
1×10^{-1}	.1 (1/10)	4.2×10^{-1}	.42 (4.2/10)
1×10^{-2}	.01 (1/100)	4.2×10^{-2}	.042 (4.2/100)
1×10^{-3}	.001 (1/1000)	4.2×10^{-3}	.0042 (4.2/1000)

The exponential system is convenient for writing very large and very small numbers, which would require many digits in conventional numbers. It doesn't take up as much room on a page, and it is easier to read and comprehend.

4.2×10^{16}	42,000,000,000,000,000
	(16 digits after the 4)
4.2×10^{-16}	.00000000000000042
	(decimal and 15 zeroes)

The exponential system is handy also for labeling charts, especially ones that have so much range of large to small numbers that they are plotted logarithmically. It is used in pocket calculators to allow a limited display area to handle a greater range of numbers than the spaces in the display would hold in conventional numbers. It is also referred to as "floating point" notation, and allows seismic recording and processing to handle a great range of amplitudes.

Binary Gain and Floating Point

The amplitudes of seismic energy go through a wide variation, from very large at the initiation of the energy to extremely small a few seconds later. When dynamite, air gun, etc., is fired, there is a loud noise, suddenly, of the explosion or pop. That is all a person can hear, as the sound immediately gets too weak for our ears to pick it up. But for seismic use we need the reflected sounds for several seconds after that. A seismic section would not be clear to our eyes if the amplitudes on it were in their original relationships. The strong energy at the first would overlap across many traces in a hopeless tangle, and the deeper reflections would mostly be indistinguishably weak wiggles on dead-looking traces.

So the early energy needs to be cut down to size, and the later energy must be amplified for easy visibility. This is gain control. An early type was the observer's hand on a knob. Starting with the gain set very low, after the shot, the observer advanced the gain manually. That is like the early movies, taken with hand-cranked cameras. The cameraman had to not get excited about the action, or he'd crank faster, slowing that action when it appeared later on the screen.

This gain control was replaced by other methods. One was a steady increase by machine, called a programmed gain control. Another system cut down the gain right after any burst of energy, and let it come back up until the next burst. That is AGC or AVC—automatic gain control or automatic volume control.

All this fiddling with the gain helped to make traces more usable, so they showed both shallow and deep reflections fairly well. But the gain controls, in making the section more uniform, destroyed the information on relative strengths of the reflections.

Binary gain is a type of gain control that works with digital recording. It is called binary because the on-or-off of the binary number system is used. A trace is recorded, starting with a certain gain. Then, at frequent intervals, the amount of energy being recorded is measured, and the gain raised or lowered as needed. A zone of either strong or weak energy is recorded at an appropriate amplitude, and a record is kept of the amount of gain that was used to bring it to that level. This record of the gain used preserves a knowledge of the absolute strength of the energy recorded. From this information, relative strengths of a reflection on successive traces can be compared. This preservation of relative amplitude makes possible the bright spot technique, which indicates gas in formations.

A gain system more advanced than binary gain is floating point recording. It uses binary numbers, but in the exponential or floating point notation, so it can handle a much greater range of numbers with the same number of bits on magnetic tape. Very large numbers are written with large exponents, very small numbers with small exponents. So the recording can include all the seismic amplitudes, from the strongest to the weakest, that are usable with present-day technology. No gain changes are required, and true amplitudes are preserved directly in the data.

Recording Format

Most seismic data is recorded on half-inch-wide magnetic tape in 9-track format, that is, with nine pieces of information in the width of the tape. The tape recorder has nine recording heads in that half inch of width. The tape is on reels, 8 or 10.5 inches across, containing 1200 or 2400 feet of tape. The reels are thoroughly protected from dirt, in individual plastic holders, which are in turn contained in sealed plastic bags. There is an older system, 7-track tape, also half-inch. Another old system uses one-inch-wide tape. It is recorded with 21 tracks of information in the inch of width.

Multiplex

Before digital recording, some of the first analog magnetic tapes were wide, maybe four inches, and were about two feet long. One tape contained the data from one shot, and had as many tracks as there were geophone groups, usually 24. The traces were positioned on the tapes like traces on monitor records, but only in magnetized form. In present-day digital recording on 9-track tapes, how can the tapes accommodate 24, 48, 96, or more traces? It is made possible by taking samples of the traces in turn, or multiplexing. The amplitude of one trace is read by the recording equipment at intervals of, for example, 2 milliseconds. Then the other traces are similarly sampled at the same interval, in between the samplings of that trace. So a sample of the first trace is recorded, then a sample of trace 2, etc. Each sample requires a number of bits (the magnetized-or-not points on the tape) to make it up. So that sample may run through all nine tracks more than once. After a sample of each of the traces has been recorded, the two milliseconds have elapsed, and another sample of trace 1 is recorded, etc.

With 9-track tape, the number representing one data sample is made by using eight of the first nine bits and also eight of the second nine. The ninth in each case is used to check the accuracy of the other eight. By using groups of bits in binary to represent numbers in octal or hexadecimal or some other code, a number as large as needed can be formed from the two groups of nine bits each. The number is a measure of the amplitude of the first seismic trace at that moment, and the sign, an indication of whether the movement of the trace is in the positive or negative direction. The next two times across the tape will have the amplitude and sign of the second trace, and so on.

There are other kinds of information on the tape in addition to the amplitude and sign of the traces. At intervals a set of nine bits contains gain information. And before any of this strictly seismic data there is considerable header information—shot point number, line number, date, and the like. Some systems use an extended header with much of the information of observer's logs—shot point location, etc.

Parity

With all those thousands of on-or-off bits of data, there are many opportunities for error. So there should be a way for a computer to do the checking, later when the data is being processed. A parity check is used for this. The eight data bits of a set of nine bits are some kind of mixture of ones and zeroes, magnetized or not. If the ones are added together, they make a number that is, of course, either even or odd. The addition is performed while the shot is being recorded in the field, and, if the sum of ones is even, another one is put in the ninth bit position. If the sum is odd, no extra one is added. So each row of nine bits across the tape should later add up to an odd number of ones. This is true all along the tape. A check can be made later by just adding up the ones in each row of digits. Any row with an even number of ones has an error in it, and can be corrected or thrown out.

Similarly, at the end of the recording of one shot there can be a parity bit on each of the tracks. Of course, the whole track won't be thrown out, but this helps to pinpoint the errors.

The field tapes are a multiplexed mixture of data from different geophone groups, and will need to be demultiplexed, sorted out into separate traces. Demultiplexing is now being done in the field on some crews. However, most demultiplexing is done in processing centers.

5

Data Processing

Improvement of the usefulness and quality of seismic data is achieved by a number of processes, some fairly simple, others more specialized or sophisticated. Corrections must be made for topographic variations and differences in LVL. Filtering is applied to retain only the best frequencies. Normal moveout is removed. Noise is reduced by stacking. Reflections are made more sharply defined by deconvolution. The data is usually taken through these and other steps in a processing center (Fig. 5-1).

Demultiplex

In the field the traces had to be recorded all together by multiplexing. But processing can best be done on separate traces. So the data is demultiplexed, rearranged into individual traces. The first sample of trace 1 is found on the field tape and copied onto another tape. Then the second sample of trace 1 is found in its position after the first samples of all the other traces, and so on until the entire several seconds of trace 1 are recorded on the new tape. If a record is 6 seconds long, and sampled every millisecond, there are 6000 samples to be copied. Sampling every 2 ms produces 3000 samples, and every 4 ms, 1500 samples. After trace 1 is complete, trace 2 is assembled and recorded on the new tape. From that tape, playouts could easily be made. They would consist of traces gathered by shots. All the traces recorded from a single shot would be on one record which would be like a monitor record or an old pre-tape paper record. For most processing uses the data will need to be in another form, combined according to common midpoint, or CDP gather.

The words multiplex and demultiplex have been shortened in processors' jargon to mux and demux.

Near-Surface Corrections

In seismic work on land, there is a problem with topography. If the ground surface is absolutely flat, then a reflection from a flat horizon would appear flat on a section, without needing any special processing.

Fig. 5-1 Processing center   Credit: PRAKLA-SEISMOS GMBH

But all traces start at zero time. So on a section in seismic time they all start at the same level, zero. But that zero time comes from geophones on the surface of the ground, which is not level. The different elevations are all plotted on a flat zero time. If there is a hill, the reflection will appear to have a lowered, synclinal shape under the hill (Fig. 5-2). A valley on the surface will produce an apparent elevation, or anticline, on reflected horizons.

The surface feature affects the data all the way down equally, as the surface is the same no matter how deep a reflection is. So the false structure can be removed by individually moving entire traces up or down the correct amount.

How much should the traces be moved? It sounds simple at first. If a hill is 40 feet high, move the traces up 40 feet, so their upper ends follow the curve of the hill instead of being at a flat upper edge of the section. But the scale of the traces is in seismic time, not feet. So some calculation must be made to convert the 40 feet into milliseconds. This is usually done in several parts, and includes the problems of LVL and the difference between the elevation of a shot in a hole or on the ground and a geophone at another location on the ground.

If the shot is fired in a hole, the sound starts some distance below ground, but is received on the surface (Fig. 5-3). To even up this lopsided situation, the time spent at the receiver end of the travel path is reduced to bring it to the level of the shot.

The time missed by the depth of the shot is measured by a geophone, the uphole jug, also called shot point seismometer or SPS,

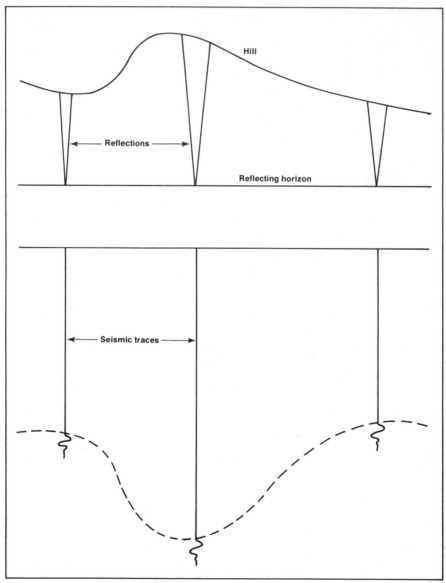

Fig. 5-2 Effect of topography

placed near—within about 10 feet of—the shot hole. The sound from the explosion is recorded by this uphole jug and the time from shot to surface measured on the record. The instant of the shot is displayed on the record as a sharp break, departure from the fairly straight path of a trace. The uphole jug makes another break on a trace, when the sound arrives at the geophone. The time between these two is measured on the record. This is the uphole time (Fig. 5-4). It is the time the sound took to go up through the material alongside the hole. This uphole time can be assumed to be the same as the time to the same depth at the receiver, so, subtracting the uphole time from the total reflection time cuts off that extra time at the receiver end (Fig. 5-5).

Shot holes are intended to be shot a little below the base of the LVL. If they are, then time spent in the LVL is also removed when the uphole time is subtracted.

If the shot is within the LVL, or if a surface source is used, corrections for the remainder of or all the LVL can be made, based on the first breaks.

First breaks are the earliest indications of energy from the shot on the various traces. The times of the first breaks after the shot are influ-

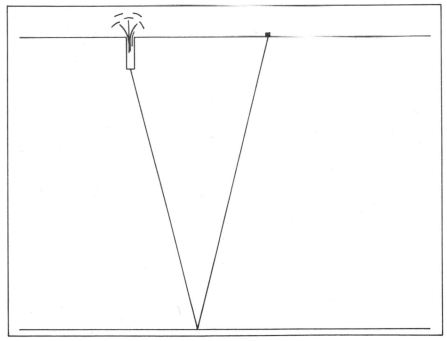

Fig. 5-3 Unequal paths

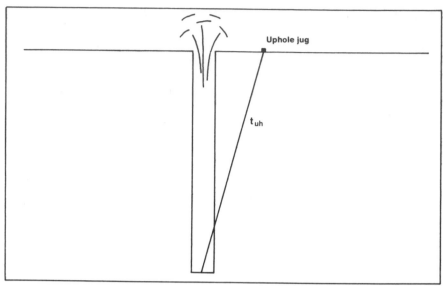

Fig. 5-4 Uphole time

enced by the distance of the geophones from the shot point and the velocity of sound through the quickest routes to the geophones. The sound goes to a fast layer, is bent along the layer by refraction, and farther along is refracted back up to the surface (Fig. 5-6). This route is quicker than a direct path through the LVL, as the slow velocity through the LVL delays the sound more than going the longer route does. So the faster route, even though longer, gets the energy to the geophones first.

Plotting these first break times on grid paper or handling them similarly in a computer program allows the time to the high-speed layer to be determined. The amount of LVL can be corrected for by subtracting time spent in the LVL (Fig. 5-7). If the first breaks are good enough, they can be picked by computer, and the first break times for different overlapping shots can be used to get a statistically best pick. This is spoken of as refraction statics (Fig. 5-8).

The datum correction is simple. A velocity is assumed for the vicinity of the level of the datum plane, and used to correct the time to the datum. If the datum is below the depth of the shot or LVL, which has already been corrected to, then the time on down to the datum and back is subtracted, removed from the travel time (Fig. 5-9). If the datum is shallower, the time is added. The reflection time remaining is time spent below the datum, so variations in this reflection time can generally be ascribed to variations in the horizons. The datum elevation is

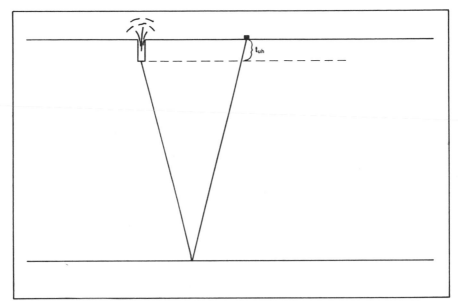

Fig. 5-5 Extra time cut off

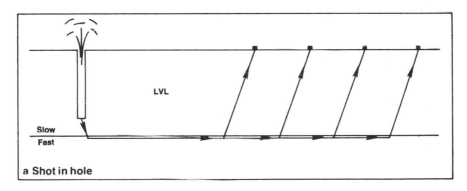

Fig. 5-6 Refraction paths

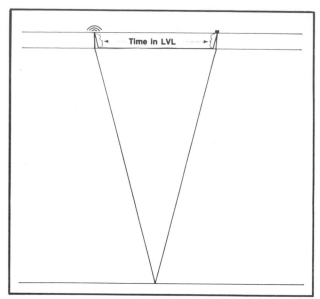

Fig. 5-7 LVL correction

chosen near the elevation of most shots, so the use of the assumed velocity is kept to a minimum. These corrections are made for each trace.

The near-surface corrections are used to move each trace up or down the number of milliseconds called for by the elevation and LVL corrections. After the traces have been moved, a playout of the data as a section may still have some irregularities that prevent the reflections from lining up cleanly, and combining well in CDP stacking.

Additional corrections can be made using the large amount of over-lapping data obtained from CDP shooting. There are a lot of energy paths from each shot, and a lot to each geophone group, so statistical analyses can be made for these points. The analyses are based on surface consistency. That is, one trace is made up of energy from one source and one receiver position. But other traces were recorded at that receiver position. All of them should have consistent corrections. Also, other traces came from the same shot. They too must agree with one another. The data can be made to yield static corrections to refine the corrections already made. Corrections to refine previously made static corrections are called residual statics. They may be made in several runs, each time refining the corrections a little more.

The type of improvement made by statics can be seen in the appearance of reflections on the final section (Fig. 5-10). A section with ragged-looking reflections that are difficult to pick can, in an optimum situation, be corrected so the reflections are smooth and clear.

100% Section

A 100% section may be made during data processing, to show how some of the steps are working. The section is made by displaying all the traces from only certain shots, enough to cover the subsurface for the length of the line only once, with no multiplicity for stack. A characteristic of this type of section is that the first breaks slant down from each of the shots displayed, giving the top of the section a saw-toothed edge. This type of display is useful mostly to the processors for quality control, QC, to check on the effect steps in processing have had on the data.

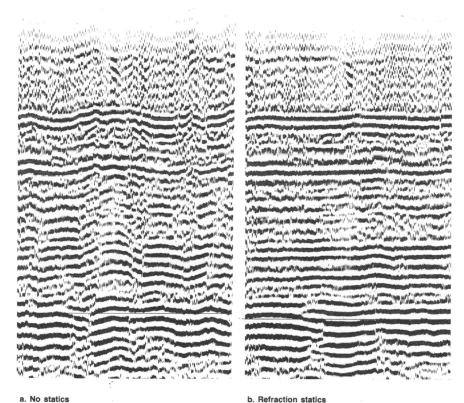

a. No statics

b. Refraction statics

Fig. 5-8 Effect of refraction statics   *Credit: Petty-Ray Geophysical Division,*
Geosource Inc.

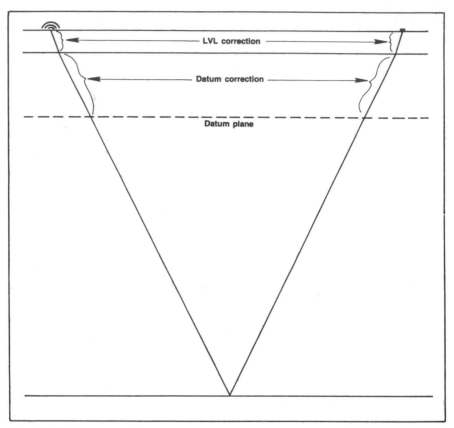

Fig. 5-9 Datum correction

Single Trace Gather

A gather is a selection of traces from among all those recorded on the magnetic tape, usually in the form of a visual display.

A single-trace gather is a display of a seismic section made up of just one trace from each shot. It may be a near-trace gather, using the trace from the group nearest the shot, for every shot, or a far-trace gather, that uses the trace from the most distant group for each shot. Or any trace between may be used, but usually the same trace for each shot, so normal moveout will not cause differences in reflection times from trace to trace. These gathers are usually made before normal moveout has been removed.

A single-trace gather then, is another type of 100% section. It is useful for giving processors an early look at data, to help in processing.

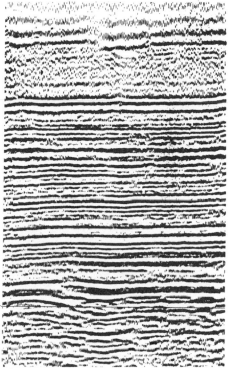

*Fig. 5-10 Residual statics applied to section
of Fig. 5-8*
*Credit: Petty-Ray Geophysical Division,
Geosource Inc.*

A look at it may show which are the stronger reflections, whether there
are steep dips or severe multiples. It can also provide a base with which
to compare a processed section for QC purposes. The reflections
should look better on a processed section, the multiples weaker.

A special-purpose single-trace gather is made on the field crew,
particularly on offshore crews, to show what kind of data is being
recorded. It is a far cry from a processed section, and doesn't have very
good data for interpreting unless the area yields extremely good data
quality, but is a useful check on the shooting.

CDP Gather

The most useful type of gather for processing is the CDP gather. It is
a display of all the traces of a common depth point, all the traces that
will later be combined to make a single stacked trace (Fig. 5-11).

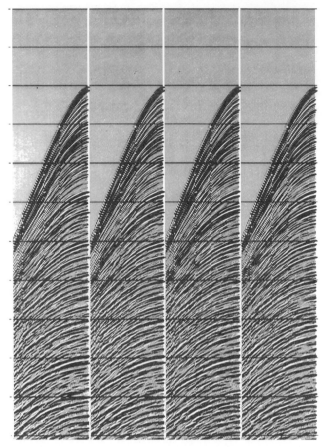

Fig. 5-11 CDP gathers from single-ended spreads
Credit: Teledyne Exploration Co.

It looks like the record of a single shot, but with some important differences. The CDP gather is a similar display of traces, but selected as the traces, not of a single shot, but of a single common depth point. These are the traces from energy that left different source positions, all went to the same point in the subsurface, and reflected to different geophone group positions. Use of the CDP gathers is correct only if the layers in the subsurface are flat, but, like other not-quite-correct procedures, it works fairly well in most cases. The gather has first breaks that slant down from near to far traces, and reflections that curve with normal moveout.

So the CDP gather has normal moveout, NMO, and can be used to determine the amount of normal moveout curvature. This is done by applying a correction to remove the NMO, and then inspecting a CDP gather or the corrected traces to see how well the NMO has been removed. If it has been correctly removed, the reflections, without NMO, should line up straight across the gather. This has application in determining velocity and in stacking data.

Velocity Analysis

The traces of a CDP gather are combined to make one stacked trace. To combine them so the reflections are enhanced, NMO must be removed. To do this, the farther traces must be pulled upward by varied amounts. That is, upper parts are pulled up more than deeper parts, to correct the different reflections' NMO curves, which become flatter with increasing depth (Fig. 5-12).

The simplest way to make this correction by computer is by trial and error. Apply different amounts of correction, and see which amount straightens a reflection.

There is a complicated equation for the relationship between velocity and NMO. Its essence is that a slower velocity produces greater NMO. With the equation, the trial and error correction can be made in the form of corrections for NMO at specific velocities.

The CDP gather can be corrected for NMO at an arbitrary velocity, say 10,000 ft/sec. The correction will not make all the reflections line up with zero NMO, but may be correct for one reflection. At that point,

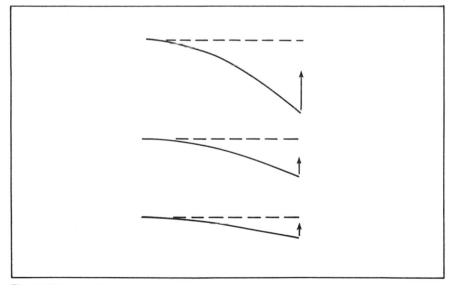

Fig. 5-12 Normal moveout corrections

the corrected reflection will have no NMO, so the reflection will be aligned straight across the gather. Reflections above it will not be corrected enough, so some NMO will remain, and the reflections will curve downward. Reflections below it will be over-corrected and will curve upward. So a look at the gather will show where that velocity applies.

If the same gather is also displayed with NMO corrected at a different velocity, like 5000 ft/sec, you can tell by inspection what seismic time 5000 ft/sec is correct for. So the gather can be corrected at a number of different velocities (Fig. 5-13), and an interpretation made to determine the velocities at the various reflections. The result can be plotted on a graph of time vs. velocity.

Exact velocities are difficult to obtain. A reflection that is almost correctly adjusted will look about as straight as one that is correct. So some other techniques have been developed to help make this fine decision. One technique is to stack all the traces of the gather, corrected at some velocity, to make one composite trace. Then the reflection that is best corrected will produce a higher-amplitude event than another. The gathers corrected with the different velocities can be stacked in the same way, and the resulting traces put side by side like a section. Looking at this display, a better determination of the best velocity can be made. Other, more sophisticated, ways of detecting the best velocity are in use. They involve computer detection of amplitudes, coherence, etc., and are displayed in various ways, with horizontal traces through the velocities (Fig. 5-14a) rather than vertical for each velocity. Or instead of a display of traces, there may be contours of the amplitudes that the traces would have had (Fig. 5-14b).

With this kind of velocity information, several things can be done. For one, a number of velocity analyses can be run at points along the line, and used for NMO corrections in stacking traces. Another use for the information is in calculating depths from reflection times. In areas of strong lateral change of velocity, this sometimes makes the difference between false and real structure.

The NMO velocity data is poorer for deep reflections, as there is less NMO to work with. But, to exactly the same degree, there is less NMO to be corrected for. So the correction works just as well at greater depths, for this one purpose of producing a good stack. But the deeper velocity data is not as good for calculating depths.

Another use for this kind of velocity data is in identifying horizons. For instance, a fault may break a horizon, and the reflection may not be recognizable enough to correlate across the fault. In this case, a certain velocity found on each side of the fault will sometimes help decide how the reflection should be correlated across the fault. Actually, the veloc-

Slow ⟶ Fast

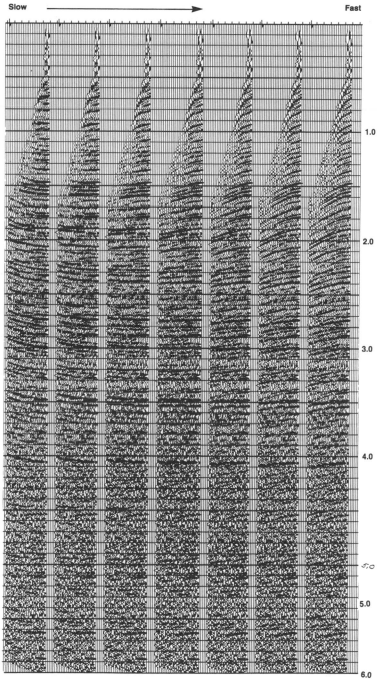

Fig. 5-13 CDP gather with NMO
corrected at different velocities

Credit: Western Geophysical Co.
of America

ities of the same formation on the two sides might not be quite the same, as greater overburden on the downthrown side would tend to make the velocity somewhat higher. And an even greater difference would probably result from the imprecision of the velocity determina-

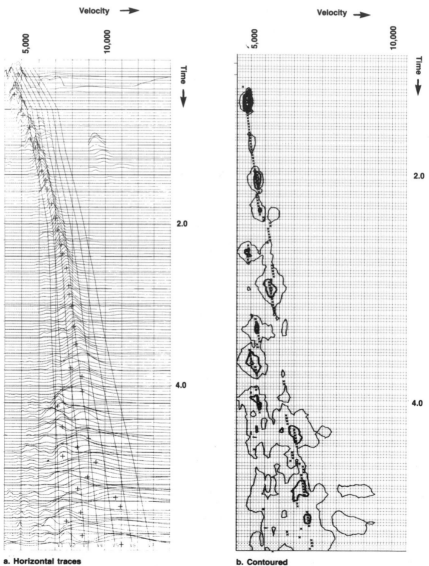

a. Horizontal traces   b. Contoured

Fig. 5-14 Velocity analysis   *Credit: **a** Western Geophysical; **b** Teledyne*

tions. A pattern of velocities might sometimes be recognizable across the fault, with abrupt changes at some horizons, etc.

This leads naturally into another use, the recognition of lithology by velocity. As an extreme example, a carbonate may stand out in contrast to a shale on the basis of velocity. And in zones of alternating sand and shale, the overall sand-shale ratio can be roughly determined by velocity.

Another use of seismically determined velocity is in the detection of high-pressure zones, to aid in planning mud programs, etc., for drilling wells.

One form of display of velocity data is as a cross section. Velocities may be plotted as time-velocity curves where the analyses are made, or lines of equal velocity can be plotted (Fig. 5-15). A popular form of velocity display is by adding colors to represent velocities to a normal section.

The commonly used name for velocities obtained from NMO data is RMS velocity. The RMS stands for root mean square, a mathematical curve-fitting method that may be used in deriving the velocities.

Velocity information can also be obtained from wells rather than from sections. And it can be determined from strengths of individual

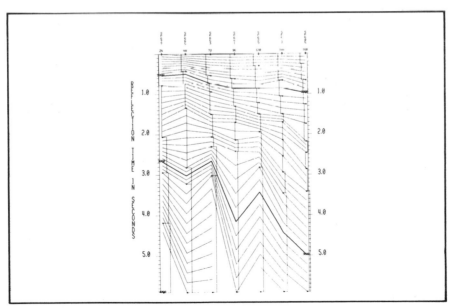

Fig. 5-15 Interval velocity curves and equal velocity lines
Credit: Geophysical Techniques, Inc.

reflections. Both of these methods are described in Chapter 7, Additional Processes.

Stacking

Stacking is the combining of two or more traces into one. This combination takes place in several ways and for several purposes.

The simplest way of combining traces is just to take the wires from two geophone groups and twist them together, so only one set of data is recorded from the two groups. This kind of blending can take place near the geophones, or at any point in the recording or processing. It can be after the energy is amplified, or after the traces are corrected for elevation differences, etc. The word "stack" is usually reserved for the combining during data processing.

In digital processing, the amplitudes of the traces are expressed as numbers, and processing is handled by arithmetical operations on the numbers. So traces can be stacked by adding the numbers together.

In either way, a peak on each of two traces will combine to make a peak as high as both added together—if the peaks are at the same times on the traces (Fig. 5-16). If they are at different times, so they don't overlap in time at all, the combination trace will have two separate peaks the sizes of the original ones (Fig. 5-17). A peak and a trough lined up will tend to cancel each other (Fig. 5-18). If their amplitudes and durations are equal, the combined trace will have no energy at that time, but will be straight.

After combining, the traces are normalized, that is, the amplitude is reduced so the doubly high peaks created by reinforcement will be of normal height. Then a peak that was on only one of the traces at a particular time, not having been reinforced, will be reduced by normalization (Fig. 5-19) . This can be accomplished by dividing the numbers of the combined trace by the number of traces combined, so the resulting trace is an arithmetical average of the traces that went into it.

Traces can be stacked for different reasons. Stacking can be a test of NMO corrections, to determine velocities in the subsurface. It can be used to combine adjacent traces, so they can be treated as one trace in processing, to reduce the amount of processing required for a large number of traces. The most common use of stacking is the combining of traces in common depth point processing. This is so well known that "stack" tends to mean "CDP stack".

For CDP stacking, a CDP gather is corrected for NMO, so the reflections on it are flat horizontal lineups, no longer curved (Fig. 5-20). The amplitudes of all the traces are added together at each sample, and

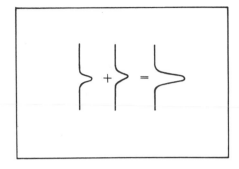

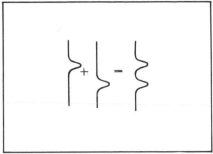

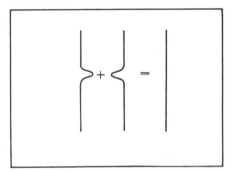

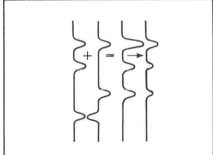

Fig. 5-18 Peak and trough combined   *Fig. 5-19 Normalized trace*

then normalized by dividing by the number of traces. That makes one CDP stacked trace. Then the next gather is stacked into a trace, and so on to make a seismic section.

CDP stack takes advantage of differences in the geometries of primary (one-bounce) and multiple reflections to weaken the multiples. To see how multiples in general are attenuated—not eliminated—suppose a primary reflection and a multiple (of a shallower primary) happen to occur at the same place on a seismic section. The primary energy went down to the horizon and back up to the surface. The multiple energy went down to a horizon about half that deep, up to the surface, down again, and up again. So the primary traverses a distance twice—down and up. The multiple covers only half that distance, but four times (Fig. 5-21). But the two used the same amount of time to travel, so they occur at the same time on the section.

In general, the velocity of sound increases with depth. Sound goes faster in the deeper rocks that are compressed by the weight of other

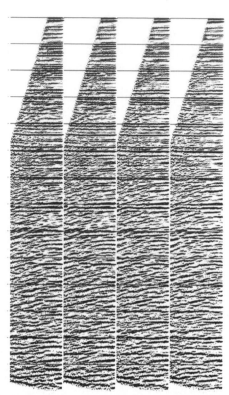

Fig. 5-20 CDP gathers of Fig. 5-11 with NMO
corrected   Credit: Teledyne Exploration Co.

rocks on top of them. So the two-way trip of the primary and the four-way trip of the multiple don't use the same velocities all the way. The lower half of the primary's course takes it through a faster zone than the multiple ever gets down to.

Now the normal moveout phenomenon on a record from the field—the curving of reflections because of the greater distance to the farther geophones—is dependent in part on velocity. The slower the velocity, the more moveout. So a multiple which stays in the slower rocks will have more moveout, more curvature, than a primary at the same time on the record.

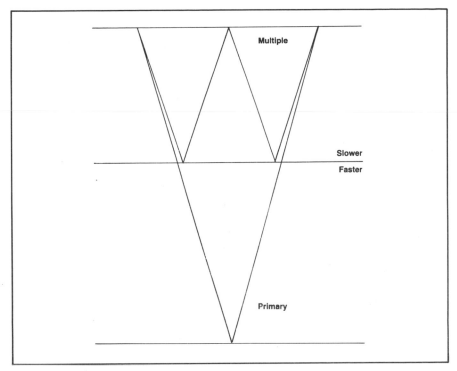

Fig. 5-21 Primary and multiple at same time

This difference is used to discriminate against multiples on sections. When corrections are made to remove NMO, velocities are part of the calculations that determine those corrections. That is, the moveout for a certain time at a certain velocity is calculated, and the traces corrected that amount to make reflections line up straight across, with no move-out. But if the velocity is wrong, the correction will not be right, and the reflection will bend up or down, overcorrected or undercorrected. The velocity, if it is correct for the primary reflection, is not correct for the multiple. To smooth out the multiple would require use of the shallow velocity. So, when the primary energy lines up straight, the multiple still has some NMO, and still curves downward.

Stacking after NMO correction for primaries weakens the multiple by combining, say, the high point of a peak on one trace with a flank of that peak on another, while the primary is combined high point to high point. This removes only part of the NMO of the multiple. When the primary is made flat, the multiple is not far from flat. So stacking won't completely cancel it. Multiples are attenuated, weakened, rather than being eliminated, by stacking.

Early in processing, before velocities are determined, a "brute stack" may be made from estimated velocities, for temporary use. This gives processors an early view of the section to help them make decisions for making the final stack, using velocities determined from NMO. Examples of brute stack are back in Figs. 5-8a, 5-8b, and 5-10.

Filtering

A seismic source sends out sound of many frequencies. Some are audible, so you can hear the pop, thump, explosion, or whatever. Seismic energy, the sound that travels well through the earth and thus carries seismic information, is only a narrow range of frequencies, from about 10 to 100 cycles per second, also called cps, Hertz, or Hz.

The frequencies that do not carry seismic information, the ones that do not penetrate far into the earth, would clutter up the section with extraneous noises—grass blades banging against the geophones, and things like that. Much of this unwanted information can be eliminated just by cutting out the frequencies that do not penetrate very well into the ground. Eliminating some frequencies is called filtering or band-pass filtering, as it allows a band of frequencies to pass, but not others.

In the old analog recording, as also in modern anti-alias filtering between geophone and digital recording, the filtering is done electrically. An electrical arrangement is used that discriminates strongly against some frequencies, and allows others to pass. The cutoff tapers down, rather than being abrupt. A filter curve shows the percentages of the different frequencies passed.

With digital processing, filtering is more exact, with cutoffs as sharp as desired. Filtering digitally is an arithmetical operation, as is any digital treatment of data. The frequencies desired can be quite precisely selected, and the result will be very close to just what is wanted. A special case is the elimination of power line frequency, 50 or 60 Hz, whichever is used in the country in which the shooting is done. In shooting on land near an electric power line, electricity can be picked up by the geophone cable by induction. To eliminate that, a notched filter is used. This is a filter especially designed to as nearly as possible eliminate just that frequency and not others. Notice that in this case the unwanted energy has entered the system as electricity, not as sound later converted to electricity.

Data processing people tend to call any operation on the data filtering, so it is sometimes necessary to specify bandpass filtering in conversation. However, the expressions "filter test", "filtered tape", and "time variant filtering" are all assumed to refer only to bandpass filtering.

Frequency Display

One of the steps in data processing is to determine what frequencies of sound to retain for the sections. A filter test, also called a frequency display or frequency scan, is made (Fig. 5-22). A short segment of section is played out several times, each time with a different narrow filter, so each playout is made up of a small range of frequencies.

The shallow part of the display is examined to see which frequencies show reflections best; then the display is examined again for deeper reflections on down the section. High frequencies do not generally penetrate deep into the earth. The higher frequency records will tend to not have many deep reflections, but will show the shallow ones in fine detail. Lower frequency records won't have the detail, but will have usable deep data. Selections are made of the frequencies that show the data best at one time on the records, the frequencies that are best at another time, etc. From tests of this sort of data from a few different parts of an area, filters can be selected for the entire program.

TVF

From the results of the frequency displays, a TVF, a time variant filter, can be applied to the data. A time variant filter is really two or more different filters applied to different zones of reflection time on the section. The bandwidths are kept fairly wide, to retain any frequencies that contribute to a reflection in those zones. The filters are usually described on the labeling of the section, either graphically or in tabular form.

Muting

In modern seismic work, the farther geophone groups are quite far from the energy source, usually 2400 feet or more. At that distance, energy reaches shallow beds not by reflection, but by being refracted along the upper part of a bed. Nearer groups do receive reflected energy. If the near and far traces were stacked together, the sound going by the two different routes would be mixed together into one trace. So, to not complicate the section with the two paths, the farther

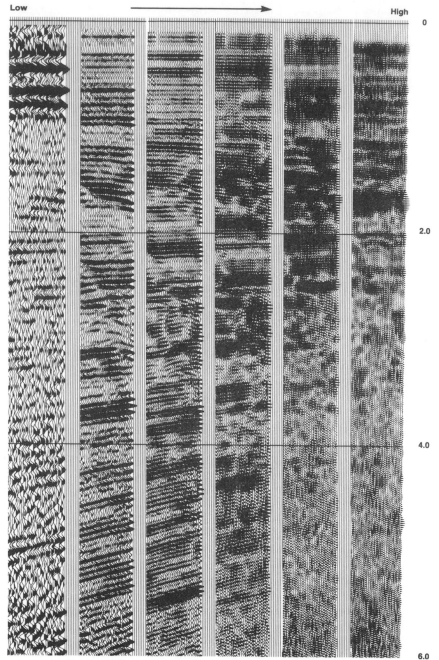

Low   High

Fig. 5-22 Frequency display   Credit: Western Geophysical Co. of America

traces are muted, cut off, down to a depth at which the traces are free of refractions. The cutoff becomes visible at one end of a section, and at any place where some of the line was not shot because of some obstacle (Fig. 5-23). The farthest trace is muted to a greater depth than the next farthest one, and so on. Within the body of the section the muting is not visible as it is at the end, or at a gap in the shooting. But above the muting depth of the farthest trace the stack is not a full stack, not the 4800% or whatever degree of CDP was shot. Often, for some shallow beds, the degree of stack may be as low as 300%.

Then, for a different reason, in offshore work the deeper parts of the nearer traces are sometimes cut off. These traces, near the boat motors, receive enough engine noise to interfere with seismic data. The shallow data has strong enough seismic energy to override the engine noise. But the deeper reflections are weak, and are drowned out by the boat noise. The farther traces are far enough from the engines to not be badly disturbed by them. So the deeper parts of near traces may be muted. This muting too can be seen on some sections.

Another reason to mute is to eliminate some low-velocity sound, like ground roll, that cuts across the data on a record. This type of sound is also low frequency, so it can usually be diminished by filter-

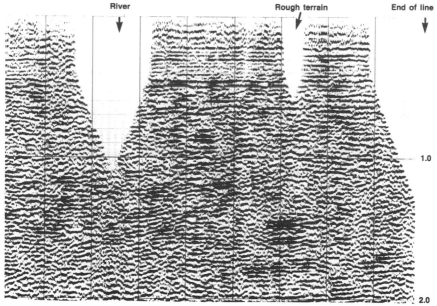

Fig. 5-23 Muting   *Credit: Petty-Ray Geophysical Division, Geosource Inc.*

ing. But if that is not sufficient, then the part of each gather that it is on can be muted out.

This use of muting, a notch mute, cuts out the parts of traces crossed by some low-velocity surface noise. This is necessary only for extremely bad noise.

Thus a "4800% section" may be 48 fold for only a medium depth, with far traces muted in the upper part of the section, and near traces muted in the deeper part. This, of course, is part of the effort to make the data as good as possible.

In connection with this, it is important, when hiring a seismic crew, to thoroughly inform the contractor as to the depths of data most necessary. Streamer length and geophone configuration should be adjusted to depth of information sought. Thus, it is not enough to say you want the best sections possible. The best possible shooting technique for a shallow sedimentary basin will not also be best for a deep basin with a deep horizon that may be prospective. Still different is a deep basin with potential pay zones scattered up and down the geologic section, which will call for a compromise technique, not quite the best for either shallow or deep data.

Signal Theory

Digital seismic processing uses techniques from signal theory, which was originally developed to improve radio and radar. It is mostly concerned with clearing up the message received, and with reducing interfering noises.

Its main techniques involve combining two sets of numbers to obtain a third list, one that bears some specific relationship to the two. This is mostly done by multiplying numbers in one list by those in the other, and is, in practical use, done by computer.

Various things can be done with this basic technique. One list can be shifted past the other and the multiplications performed at different amounts of shift. Or a list can be reversed and then shifted past the other. One list can be a specially designed list, an operator, chosen to produce a third list that is some desired modification of the one operated on.

The original lists of numbers can be numerical representations of curves, and the resulting lists can be plotted as curves. So they can be seismic traces in number form, which is the form of seismic data obtained in digital recording.

Operations using signal theory are some of the means used to improve data in digital processing. They include correlation, convolution, digital filtering.

Correlation

Correlation is one of the applications of signal theory to seismic data.

The wiggly line that is a trace on a seismic section is, in digital form, a list of numbers, so the signal theory techniques can be applied to it. The numbers are samplings of data, at intervals like every two milliseconds. The wiggles to one side or the other of the central position are represented by positive or negative numbers.

Putting two of the digital traces side by side, there are two columns of numbers. If the first number of one trace is multiplied by the first number of the other trace, then the second number by the second, etc., a column of products is the result. Adding all those products in the column together yields a single number. That number is a measure of how much alike the two traces are, in that relative position. The bigger the number, the more they are alike.

Multiplying rather than adding the pairs of numbers makes matching peaks and matching troughs both yield large positive numbers, as the product of two positive numbers is positive, and the product of two negative numbers is also positive. But a peak matched with a trough produces a negative number, as the product of a positive and a negative number is negative. So a good match is positive, and a mismatch is negative.

If one trace (list) is shifted with respect to the other and the whole process repeated, one new number, a new sum of products, is obtained. Then shift further, and get another, etc. Then list these sums of products, and plot them as a trace. This is the new trace resulting from the correlation operation. It is the correlation of the two traces.

Plotting that third list as a wiggly line trace makes a visible display of how much alike the two original traces are in the different shifted positions. There will be an amount of shift that makes the best fit, having the largest sum, the highest positive amplitude on the third trace. A shift either way from that best one produces poorer fits and smaller sums.

With the seismic traces consisting of alternating peaks and troughs, the best fit will be at a shift that puts the traces peak to peak and trough to trough. Then a shift to a peak to trough match will have a very poor correlation (negative sum) and a shift to the next peak to peak position will yield a good correlation (positive sum), but not as good as the best fit. A correlation display is a correlogram.

This, then, is a digital technique that can tell how much alike two traces are, and how they must be shifted to match best. Correlograms are curves that look somewhat like short seismic traces, with a big peak at the correct correlation, and lesser peaks at the distances the other good correlations are apart.

A special case of correlation, and at present the most useful one, is autocorrelation. As the name implies, it consists of correlating a trace with itself. That is, the two traces correlated are copies of the same one. The same procedure is used as with different traces, shifting one past the other, multiplying, adding products, listing, and plotting sums. To distinguish between the two uses of correlation, the correlation of two traces that are not the same is called crosscorrelation.

In autocorrelation, a shift in either direction from the best fit shows exactly the same features, the curve is symmetrical (Fig. 5-24). Since the two halves are identical, there is no need to show them both. So a correlogram is usually displayed by showing only one half of the curve (Fig. 5-25).

Now obviously, in autocorrelation, the best match will be when there is no shift. The traces then match perfectly. It hardly takes a computer to demonstrate that a trace matches itself when it is not shifted. But the interesting thing here is the other fairly good correlations that may show up as strong peaks on the autocorrelation trace, the autocorrelogram.

If a trace is shifted some amount, and then correlates well with itself, then it must have some feature that repeats at regular intervals. For instance, if the original trace has a strong reflection every 165 milliseconds, then, when the two copies of it are shifted 165 ms from the best fit, they will fit fairly well again. So the autocorrelogram will have a strong peak 165 ms from the best fit (Fig. 5-26).

Multiples, ghosts, and reverberations are reflections that repeat at regular intervals. The several parts of a wavelet also repeat by appearing on every reflection. So an autocorrelogram can detect them (Fig. 5-27).

Autocorrelograms, one from each trace, can be displayed side by side like a section, and can be shown with the normal section, each trace's autocorrelogram displayed above or below that trace. The display can aid in detecting multiples, etc., for processing, or in interpreting whatever remains of them after they are attenuated by processing.

Deconvolution is a process to minimize the repetitions on a trace, so autocorrelation can be used to determine the kind of deconvolution needed, and then to determine its effectiveness after it has been applied.

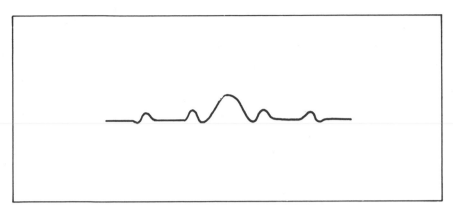

Fig. 5-24 Autocorrelogram is symmetrical

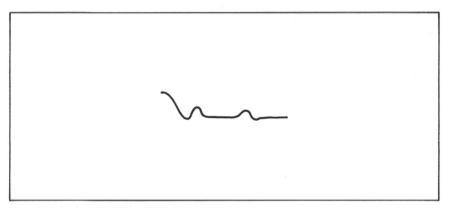

Fig. 5-25 Half of autocorrelogram

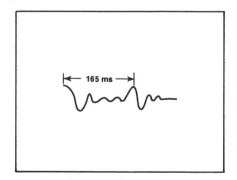

Fig. 5-26 Autocorrelogram

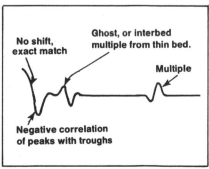

Fig. 5-27 Features of autocorrelogram

Convolution

A process similar to correlation is convolution. It is performed with two traces as lists of numbers like correlation; shifting, multiplying, adding products, and plotting sums; but it differs from correlation in that one of the traces is reversed, end for end.

The effect of convolution is not to determine how much alike two traces are, but to produce a trace that is a repeating combination of the two, the way a seismic signal combines with many velocity interfaces in the earth to produce a seismic trace. The signal is spoken of as being convolved with the earth.

To see how this works, consider a long "trace" consisting of a straight line, with a perpendicular bar at the time of each velocity interface in the ground (Fig. 5-28). The bar, a spike in signal theory language, has its length proportional to the reflectivity of the interface, that is, the amplitude of the reflection it will produce. And the spike is to the right or left, positive or negative, according to whether the velocity change is from slow to fast or fast to slow. This trace represents the velocity interfaces in the subsurface, and is put in numerical form, all points on the straight line being zero, and the spikes being single positive or negative numbers (Fig. 5-29).

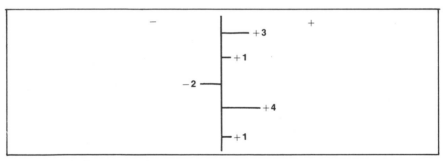

Fig. 5-28 Interfaces as spikes

Then another trace, a short one consisting of just a seismic wavelet, in the form of a list of numbers, is reversed and moved past the long trace (Fig 5-30). Assume that the first spike on the long trace is isolated, with no other spikes near it. In any position of the wavelet in passing it,

	0	0	0	0	+3	0	0	0	
	0	0	0	+1	0	0	0	0	etc.

Fig. 5-29 Spikes as numbers

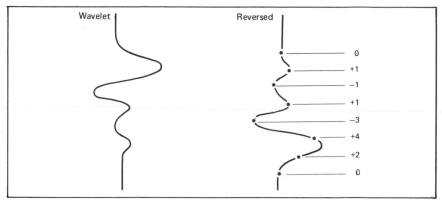

Fig. 5-30 Reversed wavelet as numbers

all the products are zero (zero multiplied by anything is zero, or none of anything is nothing), except for the one at the spike, which is the spike value times the part of the wavelet at that point (Fig. 5-31). The list of products is the product of the spike and the first point of the wavelet, the spike and the second point, etc. (Fig. 5-32). So the plot of sums of products is a plot of the wavelet (no longer reversed) multiplied by the spike. The amplitude of the wavelet on the new trace is larger or smaller, as the spike is larger or smaller. And if the spike is negative, then the wavelet is of reversed polarity, a mirror image with the left and right interchanged.

Then when the short trace arrives at the next spike, the wavelet is plotted again, larger or smaller, positive or negative, depending on that spike. And so on for all spikes. Where two spikes are close together,

First position								
0	0	0	0	0	0	+3	0	spike
+1	−1	+1	−3	+1	+4	+2	−	wavelet
0	0	0	0	0	0	+6	−	product
Second position								
0	0	0	0	0	0	+3	0	spike
−	+1	−1	+1	−3	+1	+4	+2	wavelet
−	0	0	0	0	0	+12	0	product

Fig. 5-31 Wavelet passing spike

+6	+12		−9	+3	−3	+3

Fig. 5-32 Spike-wavelet products

their wavelets will be mixed together, and not individually recognizable on the resulting trace.

In the earth, the first part of the wavelet arrives at the uppermost reflector, then the later parts arrive at that reflector. Then it arrives at the next reflector, etc. The reflections going back up arrive at the geophones in the same sequence. And the wavelets of close-together reflectors are combined into more complex shapes (Fig. 5-33).

In imitation of nature, a similar procedure is used in making a synthetic seismogram from a sonic log and a velocity survey in a well. The

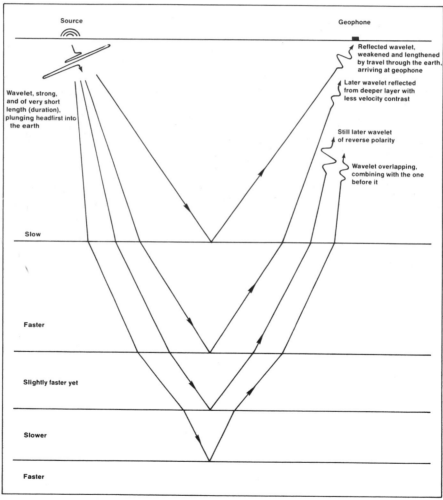

Fig. 5-33 Reflection of wavelets

well data provides information about the velocity interfaces encountered in the well. Times to velocity interfaces and characteristics of reflections at the interfaces are calculated. Then a wavelet like the one that would be produced by a seismic source is convolved (reversed and correlated) with the line and spikes. The trace formed by the convolution is an artificial, synthesized, seismic trace on which the identities of all reflections are known. It can be made more like an actual trace by adding the effects of density, and by including calculated multiples and similar features. This process is described more fully under Synthetic Seismogram, in Chapter 7, Additional Processes.

Convolution also has more general uses in processing. One of the traces convolved need not be a wavelet, but can be any trace that, in being convolved with another trace, modifies it in some desired way, operates on it. So that trace is an operator. And the other trace can be a regular seismic trace that needs to be changed in some way.

Deconvolution

When seismic energy is initiated from an explosive, air gun, water gun, or weight drop, it occurs all at once, as a near-instantaneous bang or pop. But the sound received by the geophones is stretched out into the form of a wavelet, several wiggles extending over some tens of milliseconds. A similar stretching occurs with vibrator data.

With this stretching of reflected energy, reflections interfere with others nearby. One is still going on when it is time for the next, so where they overlap, they are blended for some milliseconds.

Deconvolution is a mathematical process for partly re-compressing the stretched-out wavelet into a shorter one—not completely to a spike, but more nearly like it. The trace is processed with an operator, another trace, that, when convolved with the first, has the effect of reducing repetitions in the first one. This squeezes the different wiggles of a wavelet, so they do not extend over so many milliseconds. So a reflection is not as likely to interfere with a following one (Fig. 5-34). The reflections can be better distinguished, and more easily picked as separate events.

Ghosting and water reverberations are also repetitive, so deconvolution can reduce their effects too. Multiples that are close enough to their primaries (closer than the length of the operator) are also diminished. In doing all this, deconvolution tends to make sections made at different times and with different equipment more alike, so they can be incorporated into a single interpretation more readily.

Selection of the right operator by the processors requires fairly

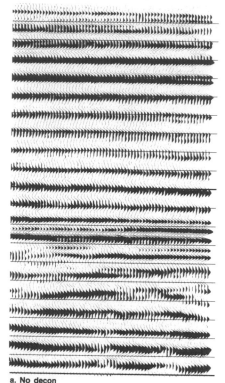

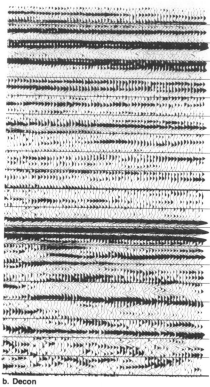

a. No decon

b. Decon

Fig. 5-34 Effect of deconvolution   *Credit: Digicon Geophysical Corp.*

good quality data, so poor records, with a lot of noise, can't be decon-volved as well. And deconvolution doesn't do anything to eliminate most noise. It works on each trace separately, but reduction of most kinds of noise calls for a method that works on continuity over several traces.

Deconvolution, or decon, is normally performed on seismic data before it is stacked, that is, on the unstacked traces. This is DBS, decon before stack. Then, after the stacking, the traces may be deconvolved again. Tests are often run to see if this second deconvolution, DAS, decon after stack, will improve the data. If it does not seem to help, then it may be omitted, to avoid unnecessary tampering with good data.

Signature deconvolution, also called wavelet processing, enhances the data still more. It can be used when the signature of the source has been recorded with each shot, pop, etc., by a geophone near the source. Or the basic wavelet for a line can be determined from the seismic data by autocorrelation. Autocorrelations can be made of parts of the traces to find the wiggles that are common to many traces in a zone of reflection times. These common wiggles are assumed to be the wavelet. This wavelet is then used to determine a deconvolution operator that fits the actual data (Fig. 5-35).

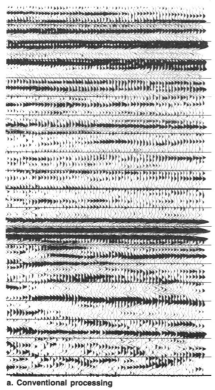

a. Conventional processing

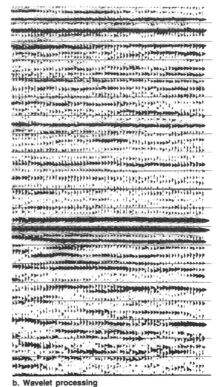

b. Wavelet processing

Fig. 5-35 Effect of wavelet processing   *Credit: Digicon Geophysical Corp.*

Fourier and Frequency Domain

If two different simple sine (or cosine) waves, each having uniform amplitude and frequency, are combined, the resulting wave is more complex than either. Adding more such waves together, a wave of any desired complexity can be obtained.

Conversely, any wave shape, like a seismic trace, can be duplicated by a combination of simple waves, using just-right frequencies, amplitudes, phases. Phase is position in seismic time, a matter of where the starting point is on the curve.

A Fourier analysis determines mathematically what waves will combine to duplicate the trace. Each of these waves can be defined completely by giving frequency, amplitude, and phase. So the seismic trace can be described by giving the frequencies, amplitudes, and phases of all the waves that make it up.

This is a different way of looking at the information in a trace. For instance, it tells more about the frequency content of a trace, and less about the time relationships in it, than a normal trace.

A normal trace is said to be in the time domain. "In the time domain" means "expressed in terms of time", that is, a seismic trace is a plot with time measured along one of the axes. It is a plot with time in one direction and amplitude in the other.

Frequency information is contained in a trace. Counting the number of wiggles in a second of seismic time gives cycles per second, but it can't just be read from one of the scales the way time and distance can.

Similarly, "in the frequency domain" means "expressed in terms of frequency". The Fourier analysis puts the trace into the frequency domain. The value of this is in having a different way of handling the data in the trace. Some data processing is better handled in the time domain, some in the frequency domain.

Frequency domain data is usually displayed as two graphs, one of frequency vs. amplitude, the other frequency vs. phase. The frequency vs. amplitude plot is useful in that it shows, in an easily understood way, the frequency content of a trace. A power spectrum is a variant of it, frequency vs. the square of amplitude. It conveys the same information, but with the differences more pronounced.

If a trace is in the frequency domain form, some frequencies can be easily and cleanly removed, by just eliminating that part of the data. It is analogous to eliminating some reflection times when in the time domain, by just omitting that part of the section. When frequency domain filtered data is put back in the time domain, it does not contain the filtered-out frequencies. This is bandpass filtering in the frequency domain.

Processing Sequence

The processing of seismic data necessarily follows a fairly uniform series of steps. Different processing centers have their own variations of processing, and may vary the order in which the steps are performed, but they all have about the same processes to go through. And some steps must be taken before some others. For instance, demultiplexing must take place before velocity analysis. The following are more or less the steps they go through, in more or less the order they use.

Processing naturally begins with the arrival in the processing center of the first batch of field tapes from the crew. They are accompanied by written records of line number, time they were shot, unusual sources of noise, field problems, etc. The same information may be magnetically recorded on the header at the front end of the tape.

The field tape is edited. It is not itself changed, but the data is recorded in modified form on another tape. It is first demultiplexed, so each trace is separate, rather than intermixed with the others as on the field tape. Dead or poor traces are eliminated. Then the traces are grouped in CDP gathers.

Tests are made to determine how much decon to use, whether to apply decon before stack only or also after stack, what filters to use, the optically best amplitude setting, etc. These tests are customarily shown to the client, and discussions held to establish how the data is to be processed. The tests include frequency displays, examples of sections with different deconvolution, etc.

The data is deconvolved, compressing each wavelet into one with less vertical extent, so the various reflections tend to not interfere with each other so much.

Statics are corrected, moving the traces up or down to correct for elevation and LVL differences.

At intervals along the lines, the data is analyzed for velocity. RMS velocities are obtained at those points.

Normal moveout is removed, using the RMS velocities at the points at which they were obtained, and interpolating between those points.

Traces are muted, eliminating shallow parts of the longer traces (farther from the shot), and perhaps deeper parts of short (near) traces.

Residual statics may be applied, straightening out irregularities left by the original static corrections.

With these corrections made, the traces are ready to be combined into one trace for each depth point. They are stacked, and a stacked tape is produced.

Sometimes it is useful to deconvolve the data again after it has been stacked.

Filtering is applied, usually as time variant filtering, in which the data for each of several ranges of times is restricted to its own set of frequencies. A TVF tape is made.

The gain recorded is taken into account in one or both of two ways, either to produce a section that has the gain so balanced as to make all

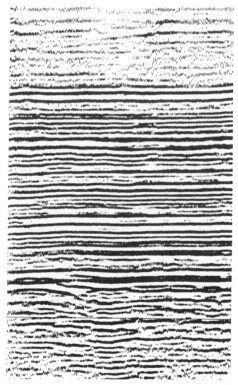

Fig. 5-36 Final stack of section of Figs. 5-8 and 5-10

Credit: Petty-Ray Geophysical Division, Geosource Inc.

reflections look fairly good, or to make a section with "true" amplitude, that is, with the strength of a wiggle on one trace in its correct proportion to the strength of one on another trace.

From the TVF tape, a section can be made on transparent film, for reproduction on paper, for interpretation. Or the data can be displayed on a screen for interpreting. A section in one form or other is the final product of normal processing (Fig. 5-36).

During the processing sections may be displayed in preliminary forms, so the processors can check them before going to final form. Also this preliminary section gives the client a chance to see what changes in the processing may be necessary.

The routine of the processors' work is filled with entering data into the computer, having the computer print data dumps, printouts usually of columns of numbers, checking them, scheduling the stages of various clients' work so they are all kept progressing and there is no time when the computer is idle.

The foregoing is normal processing that is used on most sections. But there are other types of processing that may be used to help solve the exploration problems of the area and of the client. Sections can be flattened, migrated, converted to depth. Variations of the normal processing may be used to give alternate views of the same data. Additional processing is sometimes performed as part of the sequence, sometimes done after the normal processing.

CHAPTER 6

Seismic Data Displays

All the planning, effort, and expense of shooting and processing is to acquire data and put it in a form that is useful to people in the search for oil. To accomplish this, the data is displayed in visual form on paper or on video screens. Traditionally, 2-D shooting has been displayed on paper, but displaying this information on a video screen is becoming more common. This is partly because 3-D shooting, for which video displays are ideal, has made the screens (and computers) become more available.

Vertical Sections

Seismic sections, that is, vertical sections, are the end product of 2-D shooting (Fig. 6-1). They can also be made from the closely spaced data of 3-D shooting. This type of section was described in Chapter 1 to get things started. Now it will be gone over again, this time in the context of its being produced by processing.

The recording made from a single shot does not become a distinguishable part of a section. Instead, it is taken apart into traces, and the traces modified and combined with traces from other shots to form stacked traces on the section (Fig. 6-2). The section is made up of many stacked traces. Normal moveout has been removed, so each stacked trace is in the form of a trace from a geophone close to the source, with sound going down and up nearly vertically.

The timing scale of a section is uniform. It is in seconds measured by timing lines that cross the section horizontally at ten- or fifty-millisecond intervals. Increasing time is downward, down the vertical traces. The horizontal scale though, is not necessarily uniform. Traces are spaced uniform distances apart on a section, at so many traces per inch. However, the traces represent positions on the line, and the line is assumed to be straight. But the line in the field might be curved. In offshore work, the pops might not be fired at quite the uniform time intervals intended, or the boat speed might vary. On land, physical

Fig. 6-1 Seismic section   *Credit: Western Geophysical Co. of America*

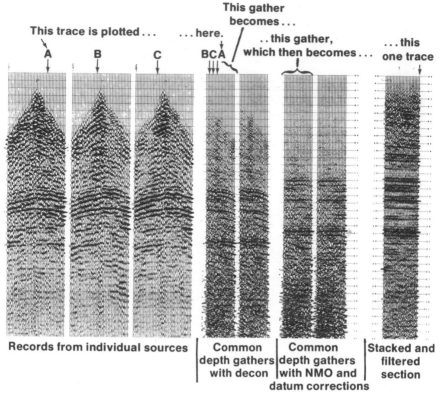

Fig. 6-2 Travels of a trace   *Credit: Houston Oil & Minerals Corp.*

obstacles or rough terrain might cause the groups to be spaced at uneven intervals. In these cases, the traces will still be uniformly spaced on the section, but those spaces will not represent equal distances on the ground. This is usually a small discrepancy. True distances between shot points can be determined by reading shot point numbers from the section, and measuring the distances between those shot points from a map, not from the section.

Offshore sections are either adjusted so the zero time on the section is at sea level, or left unadjusted so zero is at an average of the depths in the water of source and receiver, usually about 30 feet. The first reflection visible on the section will be somewhere below the zero line. This energy may represent the sea bed. However, where the bottom is soft and water saturated, there may not be much velocity contrast between water and mud, so there may be little or no reflection from that interface. In this case, the first good reflection may be from the first hard layer below the bottom.

Land data is adjusted in processing to some horizontal plane, a datum plane. In some processing, the hills and valleys appear as hills and valleys on the upper ends of the traces. In other situations, these near-surface parts of the section are cut off, so no such direct relationship can be seen. Often the surveyed elevations are plotted along the top of the section. This is useful for showing the effect that shooting on high or low places has on reflection quality in the particular area, and for making multiples more easily recognized as such. Multiples that rebound from the surface or from the LVL appear as mirror images of the topography.

Forms of Traces

Seismic sections can show the overall shapes of reflections, and also details for picking. Picking consists of marking specific points on each trace, usually bottoms of troughs or tops of peaks. Different forms of display of traces are useful in the ways they show shape and/or detail.

A seismic section on paper is most often made up of traces in the form of wiggly lines, with the peaks filled in, but there are a number of forms in which traces can be displayed. Each has its own advantages and disadvantages.

A wiggle trace section has its traces as wiggly lines alone. They are close together, and overlap where there are strong reflections. The section is fairly easy to work on at a desk, marking the lineups of reflections. Even though there is the overlap, there is no blanking out of data, so the fine detail of the wiggles can be seen best on this form. However, such a section is not so good when looked at from a distance, either so the interpreter can get an overall view, or in a meeting, to show it to a group of people. Then the traces appear to blend, so the section just looks grey all over.

A variable density, VD, section doesn't have the overlap of traces. In it, each trace is a narrow vertical band of varying shades of grey. The change from dark to light replaces the wiggle variation of peak to trough. The band is straight and of uniform width. A variable density section is easy to observe from a distance, with the reflections showing as light and dark bands across the section. But it is not so easy to handle at a desk. If the section is large scale, there can be noticeable differences in time between the parts of a dipping reflection from one trace to the next, giving it a stairstep look.

The tone range of greys cannot be reproduced with the high-contrast diazo (ammonia developed) reproduction process usually used for sections. For viewing from across a room, or for working on a small-scale section, this loss of detail is no problem, and the greater contrast may even make the image appear more distinct. But for close work, the lost detail makes precise picking impossible.

Variable area, VA, is another type of display that is easy to see from a distance. It is based on the wiggle idea, but doesn't have the complete wiggles. It consists of the peaks only, filled in solid. From a distance, reflections stand out even more clearly than they do on variable density. But, working at a desk, it is apparent that most of the detail is gone. There are no troughs, and where the peaks overlap, the solid color of one trace blanks out the tips of other traces. So neither bottoms of troughs nor tips of peaks can be picked. This too takes the precision out of picking.

A way to combine advantages and minimize disadvantages is to combine two types of display, one for precision picking, the other for visibility of reflections.

A combination of wiggle and variable density is occasionally used, with one form superimposed on the other, but it is not very popular.

Wiggle and variable area is a natural. The variable area fills in part of the wiggle, and so seems to be part of it, while making the reflections stand out. There isn't a loss of wiggle-trace detail, except where peaks overlap. The bottoms of troughs can be picked precisely. There is enough white space in the troughs for marking the section with colored pencils. So this is the predominant form of paper section, to the extent that, when a section is mentioned, people think of a wiggle-VA section.

For data shown on a video screen, the requirements are different, so the form is not usually the same as for a paper section. A paper section is interpreted by marking picks on it and reading the times of those picks from the timing lines. But in interpreting on a video screen, the precision can be obtained by the computer. Neither the exact tips of troughs nor the timing lines need be visible to the interpreter, so they are usually omitted from the display.

So, for interpreting on a screen, a variable density display may be the most convenient. These displays are usually small scale, so the stairstep effect on dipping reflections is not visible. The screen is not limited to high contrast like diazo printing, but can display many gradations between light and dark.

For some purposes, wiggle traces may sometimes be needed on a video display. Detailed investigation of reflection character can be made better on wiggle traces than on VD. So a VD display is usually used for overall looks at the data, and a wiggle form called up when needed.

Sometimes there is a reason to have more traces on a section than were obtained in the shooting. Some data processing may go more smoothly if there are more traces, or magnifying a display on an interactive system may look better with more traces. The extra traces can be added by interpolation, usually done by adding a copy of a trace beside that trace.

Polarities of Traces

A wiggle trace, with its peaks and troughs, has the peaks to the right and the troughs to the left. Whether a certain wiggle (for instance the first one of a reflection) is shown as a peak or a trough, is the polarity of the trace. If a wiggle trace is displayed the other way around, as a mirror image of the first trace, its polarity has been reversed. The two contain exactly the same information, but with different appearances. If a wiggle-VA trace is reversed in polarity, what had been peaks become troughs and what had been troughs become peaks. The new peaks—wiggles to the right—are filled in.

With the other wiggles filled in, different features of the trace are prominent. Two displays of a section may be made, one in normal polarity and one reversed. Then the two can be compared, and a feature that shows up best on one of them can be interpreted on that one.

Or the two polarities can be displayed on one section. This is often done by making the section in VA only, with the peaks of normal polarity filled in with one color and the troughs reversed so they are also peaks, in another color. For instance, there may be black peaks extending to the right, and between them, red peaks also extending to the right. People describe this as having the peaks black and the troughs red, even though the "troughs" are also in the form of peaks. Such a dual polarity section not only allows each wiggle to show up clearly, but shows effectively twice as much detail as a single polarity section.

The polarity of a trace is related to the movement of the ground when the shot was recorded. Until recently, though, the relationship was not known. An upward motion of the ground might be reversed several times in recording and processing, so that in a final section it was not clear whether that direction of motion was represented by a peak or a trough. So a "normal" polarity was just whichever polarity the sections of an area happened to have been displayed in. Modern recording and processing can keep track of the polarity, so companies have their own conventions of what they will call normal. The Society of Exploration Geophysicists has established a standard that an upward motion of the ground is to be shown as a trough. But this is opposite to the intuitive feelings that the start of a reflection should be prominent, and that a peak is like an upward movement. The convention has not been accepted by all companies. A modern section is likely to have an indication on the header as to which direction of ground motion is represented by a peak.

Vertical Exaggeration

Seismic sections do not, strictly speaking, have a ratio of horizontal to vertical scale. The horizontal scale on the section is a matter of distance, while the vertical scale is in reflection time.

However, there is certainly an apparent relationship of vertical to horizontal scale on a section. It is natural to think of both vertical and horizontal scales, on a display that is as pictorial as a seismic section, as being the same kind of thing. Thought of in this way, there is a vertical exaggeration that varies from top to bottom of the section. A tenth of a second in the shallow part of the section may represent a vertical distance of 400 feet, while the same time interval in the deeper part might cover 1000 feet.

In some respects, it would be more convenient to convert the vertical scales of all seismic sections to depth, and then make horizontal and vertical scales the same. Then all dips and fault angles could be measured by protractor. Structures would not look misleadingly steep. There would be many advantages to a geologist looking at the data.

Sections are indeed occasionally converted to depth (Fig. 6-3), and sometimes even to a 1:1 scale. But there are reasons why this is not practical for general use.

Although the determination of velocity from seismic data is becoming more accurate with improvements in data processing, it is not yet very good for determining depths to horizons. And the deeper the horizon, the less reliable the depth determination. Velocity obtained from a well is accurate at the well location. Also velocities for the same depth can vary laterally from place to place. These changes are sometimes so radical that the well velocity does not apply even a short distance from the well. So there isn't, in general, a dependable way to convert time to depth. If sections were customarily converted to depth, we would notice glaring errors of depth in many, maybe most, sections that were checked by wells drilled later.

Seismic time is the thing that is measured, and the farther we stray from using time the less definite the information is, and the less correct the conclusions from it.

As to making sections with 1:1 scale, even when the depths are reliable, it would stretch out many subtle features enough to make

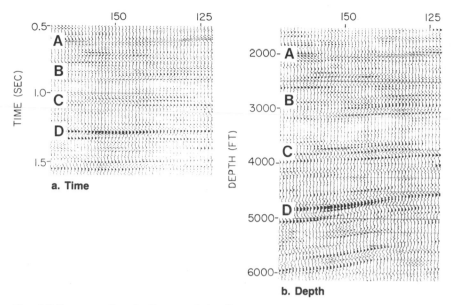

Fig. 6-3 Same section in time and depth
Credit: Western Geophysical Co. of America

them indetectable. The apparent vertical exaggeration of sections as they are normally made allows low-relief features to be seen more clearly, and so makes it more likely that they will be found. It also makes separate parts of broken reflections easier to connect.

In fact, the need for even more horizontal compression is evident in the tendency of geophysicists to look at paper sections from the end, with the eye close to the section. This produces a foreshortening that makes it possible to determine the reflection lineups a little more confidently. There is a storage and paper handling problem also. If paper sections were all stretched out horizontally to a 1:1 scale, while retaining a vertical scale large enough to show necessary detail, considerably more storage space would be needed, and more unfolding or unrolling of sections in working with them.

So conversions to depth, and to a 1:1 scale, are techniques to be used judiciously, in the cases where they will serve a special purpose for interpretation or display.

On a video screen, the scales can be varied at will, so a person can and should try different horizontal to vertical proportions to see what ratio shows particular features best. A section at a 1:1 scale, though, might be so long that very little of it could be seen on the screen at one time.

Compressed Section

A deliberately extreme vertical exaggeration for a section can be produced by making the section with its horizontal scale compressed. This can be done in processing by putting the traces closer together, and maybe combining or omitting some traces.

A compressed section is made with considerable compression, perhaps six times as compressed as the normal section, so the vertical exaggeration is extreme, and so it looks radically different from the normal section. The main function of the compression is to make subtle features easier to discern. It is used along with the normal section, to help in interpreting the regular one.

A gentle, low-relief structure that might go unnoticed on the normal section may have obvious curvature on the compressed one. Faulting that is at an apparently low angle will appear more vertical, and therefore perhaps easier to recognize. Such things as a multiple's being a mirror image of the surface topography may be clearer on a compressed section.

Just the compactness of a print of the compressed section is another of its advantages. But when using a paper print of a compressed section, reading the time of a reflection at a specific shot point or other surface point is difficult. In trying to determine a point on the reflection that is directly below a specific point on the surface, any slight deviation from the vertical by eye or straightedge may lead to a point that is too far from being under the surface point. So the regular section should be the one on which reflection times are picked for transfer to maps.

A compressed section, once made, should not be filed separately from the regular section, but attached to it. The films can be spliced together, so any print of the section will include both. This can save filing effort, and manhours in hunting around the office for sections.

A vertical section that is in the form of a display on a screen, though, can be compressed whenever it may be helpful to do so. And the problem of finding a point directly below a point on the surface is no problem on a video display. The computer can find the point correctly.

Horizontal Section

A time slice, that is, a horizontal section, is (as the latter term implies) the equivalent of a normal section, but in the horizontal direction. It is a display of the data for one seismic time from all the traces within some restricted area. A time slice is displayed on a screen or printed on paper.

Time slices are an effective way to display 3-D data. They can be made from 2-D data, but aren't very complete, with only separate lines of data. With 3-D, a time slice is like a look at the earth with the upper part cut away (Fig. 6-4). It is as though the ground had been bulldozed away down to some level, and we could see it from a helicopter. This is a radically different view. It isn't better than a vertical cut through the earth, but having both available is far better than having just one. The two show different characteristics of the earth. The throw of a fault appears clearly on a vertical section, and is difficult to deduce from slices. But a map contour can be seen directly on a slice. A time contour is a representation of the positions of a horizon at one reflection time. A slice is at one reflection time, so the reflection is a contour of itself. Another slice, at a slightly different reflection time, shows another contour of the same reflection.

Just as on a vertical section, the display consists of amplitudes of

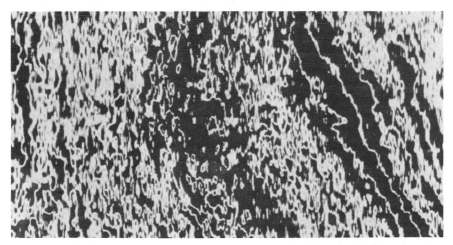

Fig. 6-4 Time slice   *Credit: Geophysical Service Inc.*

peaks and troughs. The amplitudes are shown either in variable density format, or in variable area. Variable area doesn't have the shape of a wiggle but is just a dark band, a slice through that filled-in peak. Sometimes both peaks and troughs are displayed on the same slice, in different colors, in the same way they are shown on a dual polarity vertical section.

Cubes and Other Shapes

Seismic data can be displayed as vertical sections and horizontal ones, so the two can be combined in one display. This is particularly convenient on an interactive system, but also can be printed on paper. Combining is mostly done with two displays joined at a common edge, so the eye can follow events from one to another.

One good way of looking at a solid mass of data is as a perspective view of a solid mass (Fig. 6-5). Such a display may be referred to as a cube, even though the sides are not necessarily equal in length. A view of a cube is usually made with one section shown from a nearly perpendicular viewpoint. That section shows dips, faults, etc. The top of the cube is a time slice, showing contour shapes, even though considerably distorted by the angle of view. Also distorted is the other visible

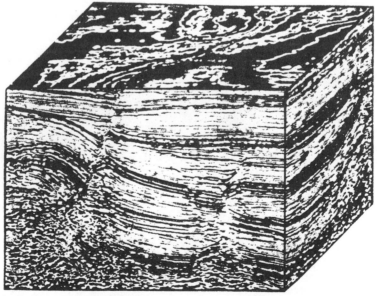

Fig. 6-5 Cube seismic display

Credit: Western Geophysical Co. of America

vertical side. But the display may not have the farther edge reduced as in a true perspective view. Of particular interest is the edge along which two planes meet. The section or slice that forms one side of the cube can be replaced with another, for instance the next parallel one. The other sides can automatically be extended or shortened to meet the new side. This process can be continued, so the view progresses through the cube. On some systems, the cube can be rotated, so it can be seen from different angles.

Another convenient semi-perspective display is a "chair", consisting of a slice with vertical sections rising from its far edges, and others dropping down from its near edges (Fig. 6-6).

A useful non-perspective display is of a slice with four vertical sections extending outward from it on four sides (Fig. 6-7).

Any of these displays can be changed progressively from section to section or slice to slice through a volume of data in the same way as a cube is changed.

Color

When color displays of seismic data began to be used, they were made in the form of photographic color prints. This method is still used

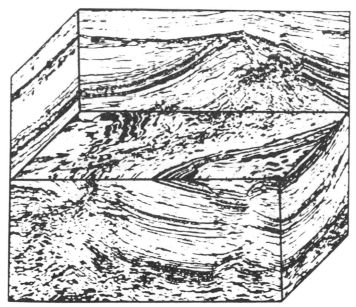

Fig. 6-6 Chair display

Credit: Western Geophysical Co. of America

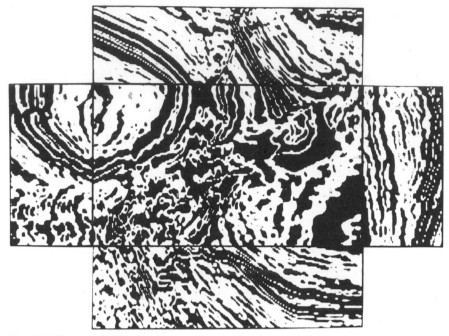

Fig. 6-7 Slice and four vertical sections
                    Credit: Western Geophysical Co. of America

to some extent. However, 3-D produces such quantities of data that it was necesssary to develop faster and cheaper ways to make large color prints. Color xerography, given an image by computer, is one way of making color prints rapidly. Another principle, the ink-jet method, uses a machine that, under computer control, squirts tiny jets of colored inks onto plain paper to make a printout. A large display can be made in a few minutes. Ink-jet printers are made in various sizes, from printing on letter-size paper to about drafting table size. Most use rolls of paper, so in one direction they are limited only by the length of a roll of paper. The printed-out 3-D displays that are most useful, though, are not much longer than they are wide.

Color sections may be normal wiggle-V.A. or V.D. sections, with some parameter added in color. Velocities may be shown as gradations from red through the spectrum to blue. Frequencies or polarities may be shown in color, maybe by colors selected to contrast with adjacent colors. Even amplitude or dip, already visible on the section, may be given added stress with a range of superimposed colors.

The color sections put more data in usable form on the sections, for direct comparison with the traces. For instance, a person can see and think about velocity while looking at structure, and better understand the interrelationships between the two than by having them on separate displays.

On a video screen, colors can be selected to suit the immediate purpose. The interpreter can vary the color scheme of a display until it appears to show some feature best. Whatever display is chosen, it can then be reproduced on a color printer.

Thus, even an interpreter who is colorblind to certain colors, for instance, unable to tell red from green, can select an appropriate color display to work with. Another interpreter, with different color perception, could readily change the colors as desired. The remaining problem, though, is that a printed color display might lose some of its effectiveness when shown at a meeting in which someone was colorblind. But most displays have enough variety of colors to show the data to anyone.

CHAPTER 7

Additional Processes

In addition to the data processing that produces a stacked section, there are other processes that are applied to some or all of the data. Reflections are migrated to more correct positions, other displays and processes are applied, velocity information is obtained in other ways and used for other purposes.

Migration of Dips

A vertical seismic section represents a cross section of the earth. It appears most like the real earth when layers are flat, and fairly much like it when they have gentle dip. With steeper dip, the reflections are not where they are in the earth, and have different amounts of dip. Dip migration, or just migration, is the process of moving the reflections to their proper places, with their correct amounts of dip (Figs. 7-1 and 7-2).

When energy is recorded from a geophone near the shot point, the energy has gone almost straight down to the reflecting horizon and back up—if the reflector is flat and velocities are uniform (Fig. 7-3). In CDP stack, all the traces are corrected for NMO, making them look like traces recorded by geophones at the source positions. If the horizon dips, the energy goes to and from it by the most direct route, meeting the reflector perpendicularly. Energy that hits at other angles will go off in other directions, and not be received by that geophone (Fig. 7-4).

This perpendicular reflection is the basic idea behind dip migration. All the rest of migration follows from the perpendicular reflection principle. Structure and velocity variations cause the sound to follow non-straight paths down to the horizon and back up, but, right at the reflecting surface, the energy path is perpendicular to that surface.

So the reflection point is not under the shot point, but offset from it. And the geometry makes it be offset in the updip direction, up the slope of the bed.

That is what happens to the reflected sound. But how does it appear on a seismic section? The traces on a section are parallel. They all hang straight down from the surface, because they are just measurements of

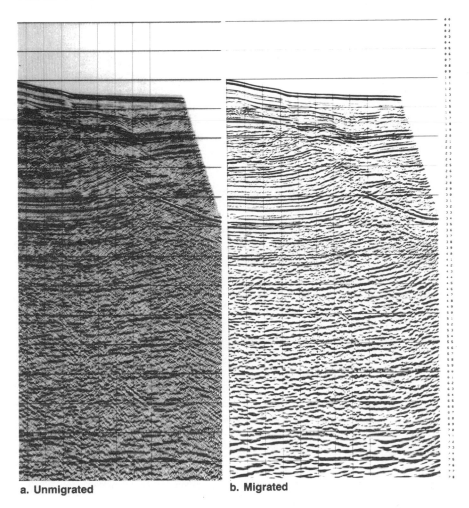

a. Unmigrated

b. Migrated

Fig. 7-1 Effect of migration   *Credit: Teledyne Exploration Co.*

the times for a sequence of sounds to strike the geophones. So the time during which sound traveled at a slant is shown on the section as straight down (Fig. 7-5).

Using the perpendicular reflection principle, some subsurface features and how they will look when converted to sections with vertical traces can be considered. Then some rules can be formed for how the features of the section will have to change to be migrated back to their correct configurations. For simplicity at this stage, it will be assumed that the velocity of sound is constant all through the geologic section,

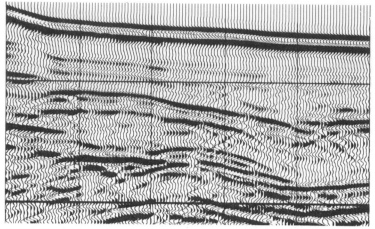

a. Unmigrated

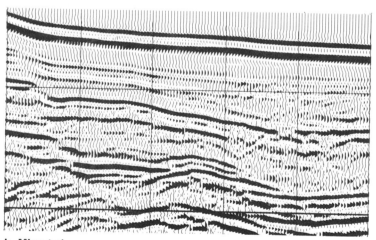

b. Migrated

Fig. 7-2 Detail of Fig. 7-1   *Credit: Teledyne Exploration Co.*

and that the lines are shot in the direction of dip, so they do not have any reflections from one side or the other of the line.

1. The ray paths—routes the sound travels—are at an angle, and when displayed on a section as traces, change to hanging vertically. So migra-

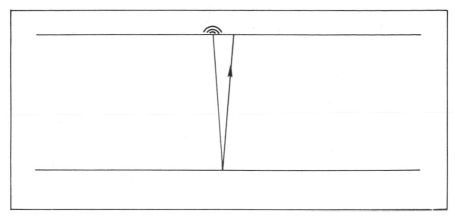

Fig. 7-3 Near-vertical reflection

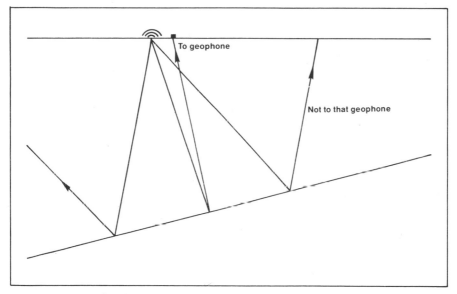

Fig. 7-4 Dipping horizon

tion must move them back updip (Fig. 7-6). Therefore, in migration: *Reflections move updip.*

2. That is just dip in one direction. Now picture an anticline. In the subsurface, the ray paths reach inward to reflect perpendicularly from the horizon. Then, on a section, they swing down from the vertical, becoming more spread out, so the feature looks broader. So, migration

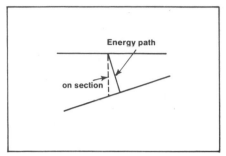

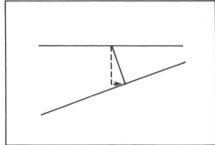

Fig. 7-5 Vertical trace   *Fig. 7-6 Migration moves data undip*

back to the correct position will narrow the structure. In migration:
Anticlines become narrower.
3. Also in the anticline, the traces swinging down to the vertical position show more relief. So migration will reduce the amount of relief (Fig. 7-7). However, this is true only if the anticline is isolated on the section. But if dips change progressively to flat at the flanks, the total relief will not change, as the flat parts are not affected by migration. Migrating:
Anticlines may have less or the same vertical closure.
4. The very crest of the anticline doesn't move. Right at the top, there

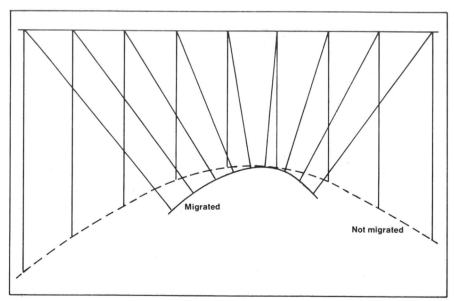

Fig. 7-7 Migration narrows anticline and reduces dip of isolated anticline

isn't any dip, so that part won't go anywhere. The energy path there is vertical, so doesn't undergo any change in being put in section form. So:

The crest of an anticline does not move.

5. In a syncline, the ray paths have to reach outward to be reflected perpendicularly. On a section, they make a narrower-looking feature. So migration must spread it out again (Fig. 7-8). When migrated: *Synclines become broader.*

6. And like the anticline's top, the lowest point of a syncline doesn't have any dip, so energy goes straight down to it, and isn't changed on a section or in being migrated. Migrating:

The low point of a syncline does not move.

7. The amount of relief of a syncline usually isn't very important to exploration, as it is the anticlines that have the oil. It is, though, an inversion of the anticline situation. Migrated:

Synclines may have more or the same closure.

8. Synclines can behave in another way on sections, though, if they are relatively deep in the section or narrow. The deeper or sharper ones have ray paths that cross on the way down, with one trace being in a position to receive information from two or even three parts of the syncline. Two crossing lineups of energy on a section, with an apparent anticline perhaps visible beneath them, will be on the section (Fig. 7-9 and 7-10). This is spoken of as a buried focus, as the seismic energy crosses somewhat as light rays do when focused by a lens. And, from its

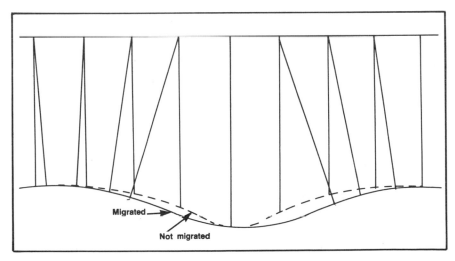

Fig. 7-8 Syncline becomes broader

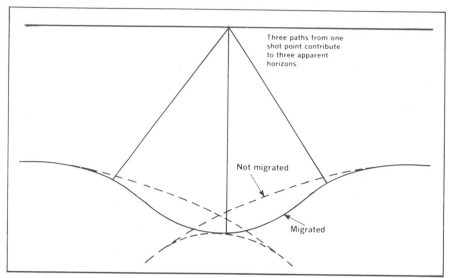

Three paths from one
shot point contribute
to three apparent
horizons.

Not migrated

Migrated

Fig. 7-9 Crossing energy paths

appearance on the section, it is also called a bow tie (Fig. 7-11).
Migrated:
Crossing reflections may become a sharp syncline.
9. Now there is a still more involved situation. Where a fault or other
sharp change breaks the continuity of a horizon, the point of change
returns energy to any source within range, in any direction (Fig. 7-12).
That is, it behaves as a new source of energy, or, looked at another way,
as a tiny effectively spherical reflector that is perpendicular to every ray
path. So energy is returned to a number of shot points at different
distances from it. Then on a seismic section, with the reflected energy
vertically below the shot points, it will become an apparent anticline. It
is a very regular one though, like an upright, open umbrella. The form
is actually a hyperbola. This point-reflector situation is a diffraction, not
to be confused with a refraction. Diffraction is generally recognized
easily by its regular shape. Sometimes only half of it is visible, so the
broken-off formation appears to continue in a smooth curve downward
(Figs. 7-13 and -14). Even though it isn't a normal reflection, a diffraction
pattern is created by the traces hanging straight down, so the same
process of migration applies, moving the ray paths to their correct positions. In migrating the diffraction, they fall back to a point. When
migrated:
An umbrella shape, diffraction, becomes a point.

10. Also, the diffraction pattern has no dip at its crest, and the crest is the point where the diffraction occurred. The crest is at the isolated point that produces the diffraction. Migrating:
The crest of a diffraction does not move, and is the diffraction point.

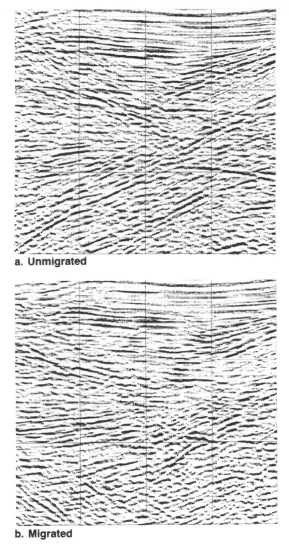

a. Unmigrated

b. Migrated

Fig. 7-10 Buried focus

Credit: Petty-Ray Geophysical Division,
Geosource Inc.

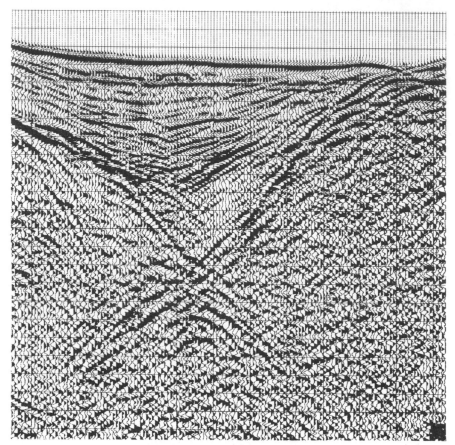

Fig. 7-11 Bow tie   *Credit: Western Geophysical Co. of America*

Dip Lines and Others

All the migrating described so far has been in two dimensions—along a 2-D section. This is fine if the section happens to be aligned along dip, perpendicular to strike, not so good otherwise. But there are a lot of reasons why the line may be run in some direction other than along dip:

Terrain may dictate the direction of the line,

The line may run from one well to another,

It may be a tie line between other lines,

The direction of dip may not be known before shooting—except regionally,

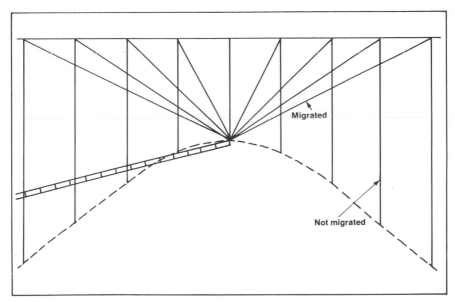

Fig. 7-12 Point in subsurface

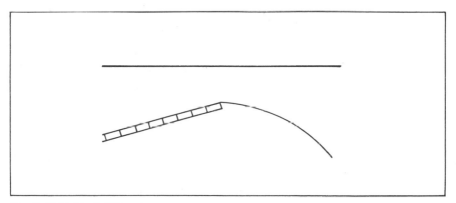

Fig. 7-13 Half of diffraction

Different horizons may dip in different directions so a line that is downdip on one horizon is not downdip on another.

If the line isn't a dip line, then migration along the line isn't correct. Updip—the direction reflections migrate—isn't in the plane of the section, vertically below the line. The 2-D migration doesn't take care of this effect.

Consider the influence that direction of shooting has on straight lines over a bed with uniform dip all in one direction, a monocline. If a

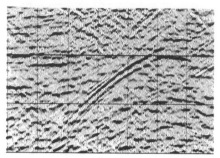   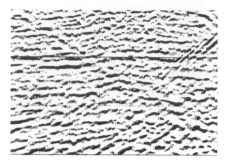

a. Unmigrated   **b. Migrated**

Fig. 7-14 Half diffraction   *Credit: Teledyne Exploration Co.*

line is shot along dip, all of the dip is represented on the section (Fig. 7-15). If, however, the line is shot along strike, no dip at all is recorded (Fig. 7-16). The horizon looks flat on the section. Anywhere between the two, the section shows some dip, but less than the actual amount (Fig. 7-17).

Or a seismic line over the monocline may be curved. A line is laid out as a straight line if at all possible, so the data can be stacked properly. However, there are occasions when a line must curve or not be shot at all. Examples are marine-type shooting in a river, which must follow the river's course, and land shooting along a narrow winding ridge or valley. These curved lines constitute a special situation if there is also steep dip in the area. If a curved line wanders up the uniform dip and then down, an apparent structure is formed on the section (Fig. 7-18). It may look geologically reasonable, with less dip in the shallow beds than the deep. Migration of that line can be extremely misleading.

If the line had been shot as two straight lines, with the bend skipped, the reflections on each would move updip on the sections. But, if the curved line is migrated in two dimensions, what happens? The segments of reflections are migrated around the bend. Now, some of the dips are turned around, and the section can seem to have a syncline if the bend is sharp enough.

That is an extreme case, with its hairpin turn, and might never happen. It illustrates the point, though, and less sharp bends produce some of the same type of confusion. They make dips do confusing and incorrect things. The moral of this is just that sections can't be safely migrated without thinking of special factors that may influence them.

Many seismic horizons, of course, aren't monoclines. There can be dips in all directions. And dip from different directions can be recorded on the line. Data that is recorded, not from under the line, but from points in an updip direction to the side of the line, is called side-swipe. It is most noticeable when it crosses other dips on the section. The crossing dips occur in complicated geology, where there are updip reflectors in different directions from the line.

Migration Methods

Migration is done in several different ways in data processing. There are three main ways, and several variants of each of the three. The Kirchhoff summation method combines data found along curves on the

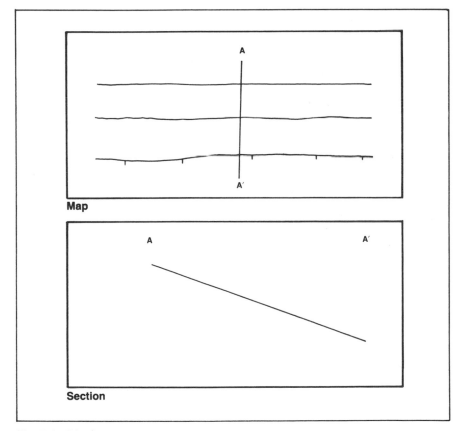

Fig. 7-15 Dip line

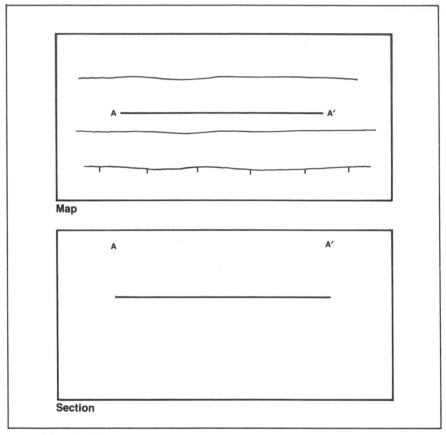

Fig. 7-16 Strike line

section. Finite difference migrates layer by layer. Fourier transform migrates data that is expressed in terms of frequency and wavenumber. Each has its own advantages and disadvantages, making it apply best to certain situations. A processor or an oil company then, does not normally have one preferred method, but uses whichever method or variation seems best for the particular application.

The first of the three, Kirchhoff summation, derived from earlier hand migration. Migration by hand was based on the idea that seismic energy reflects perpendicularly at the reflecting horizon. So geometrical devices were thought up to make the angle down at the reflection point be a right angle. If you have two traces hanging straight down to a reflecting surface that dips, the traces do not meet the horizon at a right

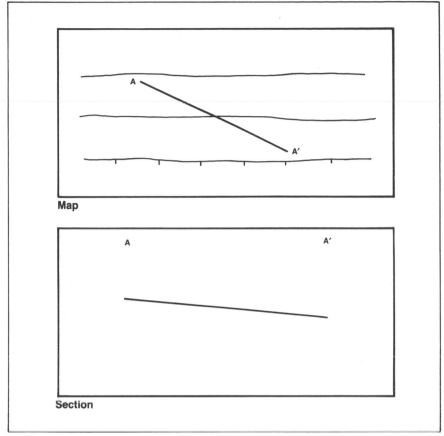

Fig. 7-17 Intermediate line

angle. But they can be swung from pivots at the top of the section, until they do form right angles with the horizon—the line connecting them. The reflection is then in its migrated position—if the velocity of sound is totally uniform down to the reflecting horizon.

Computer migration at first was an extension of the arc-swinging of hand migration. A trace was pivoted in an arc, and the data on it recorded in every position of the arc. The next trace was similarly handled, and so on, for the entire section. All these overlapping traces were added together along vertical lines to form new traces. At most places on such a combined mass of data, peaks and troughs would add together at random, largely cancelling one another. But a reflection, at

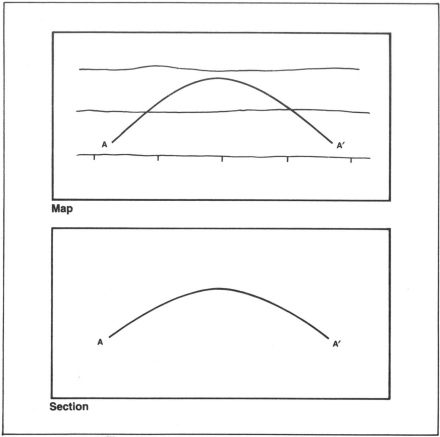

Fig. 7-18 Curving line

its migrated position, would be an accumulation of energy, peak to peak and trough to trough, so it would be reinforced. The result of the computer accumulation was a migrated section.

Another hand method used a diffraction chart, a plot of the diffractions that would be formed by many isolated points in the subsurface, one above the other. A diffraction on the section that matched one on the chart would be migrated by re-plotting it at the crest of the one on the chart. Just a piece of a diffraction could also be matched to one on the chart, and plotted at its top. Similarly, a segment of a reflection, matched to part of a diffraction, could be migrated by plotting it at the top of the diffraction curve. A computer version of this method, called diffraction collapse, was accomplished by stacking together all the data found on a diffraction curve, at the top of that curve. Doing that for every diffraction, at every trace position, produced a migrated section.

The arcs or the diffraction curves can be made with distortions that allow for velocity variations with depth. These two methods are mathematically equivalent, and are also similar to a formula, the Kirchhoff integral, which can be applied to data to migrate it. These related techniques are called Kirchhoff summation or some similar name.

A minor disadvantage of a section migrated by summation is the presence of noise arcs. These are alignments of seismic energy in the form of parts of circles, open side up, so they are also called smiles. They occur mostly in the deeper part of the section, where there are not many reflections, so random noise is free to create these artificial shapes. They give the bottom of the section a scalloped look. Smiles also appear near the end of a section, where the absence of data beyond the section makes the summation incomplete. They are not much problem, as they can easily be recognized and ignored. They don't interfere much with reflections, as the smiles are composed of weak energy that is easily overridden by reflections.

A greater disadvantage of some summation methods is that they do not preserve the fine detail of reflections very well, probably because of the blending of data in combining across arcs or diffraction curves. But the Kirchhoff integral can be manipulated to preserve that detail. The method has the advantage of handling steep dip, over 45 degrees, better than other methods. It also handles shallow data better, because muting can be applied before the migration.

Finite difference migration is another method used by data processors. It takes an entirely different approach, using a formula that extends a field of data back to an earlier stage. So, from the data received at the geophones, an entire new set of data is calculated—the data as it was just a little time before it arrived at the geophones. In effect, though not usually plotted, it is another seismic section, the one that would have been recorded if the geophones were not on the top of the ground, but buried a little way down. Then another section, in effect a little deeper, is calculated, and so on, all the way down to the reflecting horizons, and on to the bottom of the section.

As data is retraced, things are put in adjusted directional relationships. That is, they are migrated. It is somewhat as if you just migrated the shallowest reflection on the section, then used it as a basis for migrating the next one, etc. The steps are about 20-100 ms long, depending on the degree of dip to be migrated. Steeper dips require shorter steps, and therefore more of them. This makes the process slower and more expensive. At each stage the data above has been migrated and the data below is not yet migrated.

The resulting migrated section retains the reflection character, the distinctive appearance of the reflections, of the original section. This non-degradation of the reflection quality of the section is one of the main advantages of this migration method (Fig. 7-19). The finite difference method excels in its handling of lateral velocity variations. A section migrated by finite difference shows detail as well as an unmigrated section does. It is good in the deeper part of the section, as it does not form smiles as much as other methods do.

A disadvantage of finite difference migration is that it does not, in its pure form, handle steep dips well. So various modifications have been developed, giving it some of the steep dip handling characteristics of diffraction collapse migration.

The third method is Fourier transform migration, f-k migration, frequency-wavenumber migration, or some similar name. It is accomplished by expressing the data in terms of frequency and wavenumber rather than time and space, and re-arranging it into migrated position.

Sections can be thought of as plots of reflection time against distance—time plotted downward, distance to the right or left. The data is said to be in the time domain. Being in a domain means that it is expressed in terms of that thing. The information can as well be put in the frequency domain, plotted in terms of frequency against horizontal wavenumber. Wavenumber is the reciprocal of the wavelength. The data is changed from being in terms of time and distance in two Fourier transforms. The first replaces time with frequency. The second replaces distance with wavenumber. Then in this new form, the data is migrated.

Dips are in effect gathered together, all the one-degree dips at a certain depth in one place, all the two-degree dips at that depth in another, etc. So if a one-degree dip at 1000 feet should be migrated a certain distance, all the one-degree dips at 1000 feet are migrated that distance, no matter where they are along the section. Since depths, and not times, are used in this process, then to use this method the data must first be converted to depth. To apply varying velocities to it, the depths must be changed to simulate the velocity differences.

After the dips have been moved, the data is converted back, again in two steps, to the time domain, and all the one-degree dips at 1000 feet go to their proper places along the line. This distribution occurs in a sort of wholesale manner, and therefore has its own shortcomings. If a dip migrates to a position beyond the end of the section, then the literal-minded conversion happily finds a place to put it—at the other end of the section.

The way to prevent this wraparound is to add dead traces—traces that are straight lines—at each end. Plenty of traces are added, so there

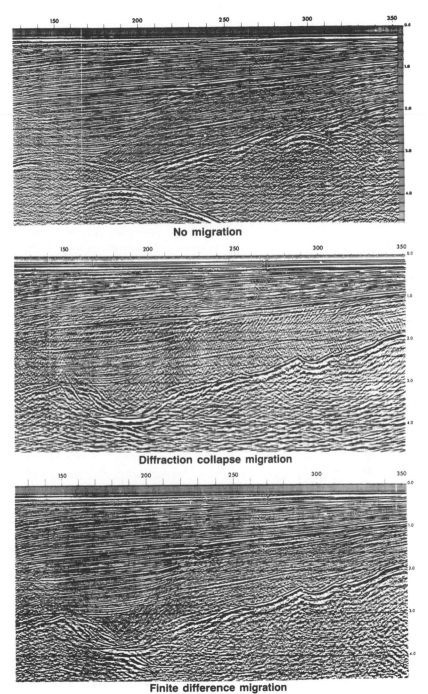

No migration

Diffraction collapse migration

Finite difference migration

Fig. 7-19 Two migration methods *Credit: Seiscom Delta United*

is room for any dips to migrate out there. Then when a section is played out on film or video screen, those blank traces are omitted. Wraparound also occurs between top and bottom of the section. It is taken care of by adding extra blank length to the ends of the traces. This extra length also is left off of displays.

Advantages of this method are that it is the fastest, therefore the cheapest, and that it is a method that preserves the character of reflections.

There are variations and combinations of these three main methods, thought up to reduce their disadvantages. So a data processing company may have one of the methods in a special form that is largely free of the problems that method usually has. The type of migration that should be used depends on the area and the goal in the seismic investigation—whether there is much dip, whether there is much velocity variation, whether the horizons of interest are deep or shallow.

Today, almost all seismic data is migrated. Migration almost always improves the interpretability of data. Faults are more sharply defined, diffractions are eliminated or reduced, sharp synclines are cleared up. The improvement occurs with any method of migrating. It occurs even when the wrong velocity is used, as long as it is somewhere near the correct one. It occurs when strike lines are migrated. It occurs when there is not much dip. The migration is better when the appropriate method is used and the velocity is correct. It shows more useful data when lines in the dip direction are migrated. And lines with significant dip have more to migrate. But almost any time data is migrated, it is improved.

Migration is not dependent on having CDP data. It can be performed as readily on 100% data. Without the stacking of CDP data, the final result is not as good, but the improvement caused by migration is just as great as with CDP data.

Migration Before Stack

Migration is customarily done after stack, for economic reasons (Fig.7-20). Migration done first and then followed by stack yields better sections (Fig. 7-21), but for 96-fold shooting, there are ninety-six times as many traces to migrate. But for special reasons, it is sometimes worth the price to migrate before stack.

In CDP shooting it is usually assumed that sound is reflected at the midpoint between source and receiver. That would make all the traces, with different source-to-receiver distances, reflect from the same

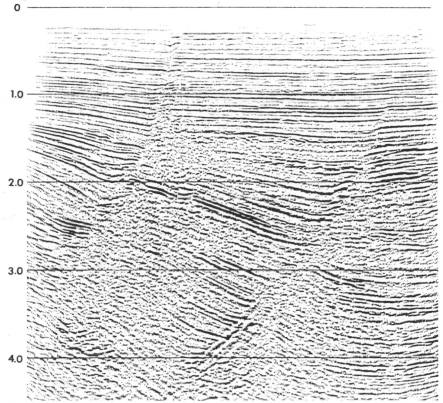

Fig. 7-20 Migration after stack   *Credit: Digicon Geophysical Corp.*

place. But they do reflect from the same place only if the reflecting layer is flat. If a reflecting horizon dips, then the reflection points for different source-to-receiver distances are different. The traces are usually stacked anyway, without causing much smear of the data. But for the best migration, the traces should be migrated in separate sections—near-trace section, second-trace section, etc.—and then stacked after each trace has been migrated to its correct position.

There are some techniques for simulating the effects of migrating before stack without migrating so many traces. One of them, dip moveout migration, applies a correction for dip to data before stack. It is sort of a partial migration before stack. The correction removes the smear caused by different reflection paths not being reflected from the same point. Then the data can be migrated after stack, with results about as good as if they were migrated before stack.

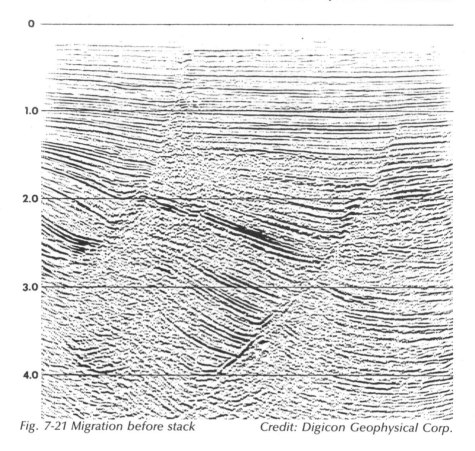

Fig. 7-21 Migration before stack   *Credit: Digicon Geophysical Corp.*

Three Dimensional Migration

When an area is shot with a 3-D technique, it can be migrated in three dimensions. This eliminates the problem of side-swipe. Two dimensional migration necessarily leaves side-swipe on the section. But migrating in 3-D moves data to its proper position in any direction. So a section made after 3-D migration has only the reflections that are actually from directly below the location of the line on the ground (Fig. 7-22).

Migration in 3-D is done using the same principles as are used in 2-D. But to swing arcs in all directions, or correct layers over an area, etc., uses many times the computer time that is needed to do the same thing along a single line. One way to handle it is to use an especially powerful computer. This is expensive, and calls for sending the data to some central location. The usual way around this is to migrate the data

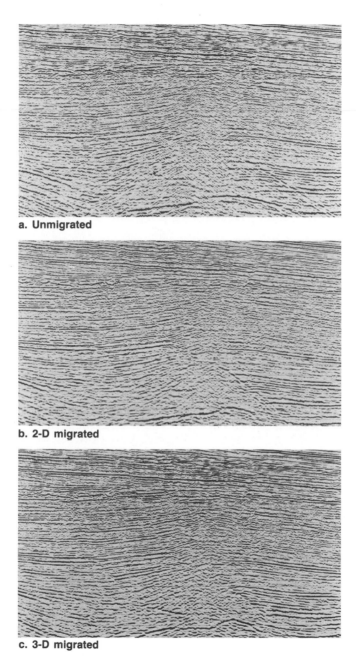

a. Unmigrated

b. 2-D migrated

c. 3-D migrated

Fig. 7-22 Effect of 3-D migration

Credit: Geophysical Service Inc.

in one direction, as a collection of parallel 2-D lines, and then to migrate that migrated data in the direction at right angles to the first lines. This is considerably quicker and cheaper, and the results are nearly as good. Most 3-D data is migrated in this two-step manner.

True Amplitude

Seismic data is routinely recorded in floating point form (Chapter 4), retaining a record of the exact amplitudes of the energy. This precision can be used to produce a true amplitude section, which has true relative amplitudes from trace to trace. Necessarily, the shallow data is restrained and the deeper is built up to keep both kinds of information visible on the section, but the increase is made at about the rate that they diminish as they get deeper in the earth, so amplitudes are also comparable vertically.

A true amplitude section (Fig. 7-23) generally looks paler than a normal section (one with some kind of gain control). The true amplitude section has more difference between strong and weak reflections. It is not as easy to pick, because any weak reflections or weak parts of strong reflections are shown with low amplitude. Its function is to show amplitude variations, for evaluating amplitude anomalies. Its use for this purpose is to compare strengths of reflections, as discussed in Chapter 9.

Flattening

There are some advantages to moving the traces on a section up or down so that one reflection becomes flat. One purpose of flattening is to get around near-surface problems and produce a section that shows reflections and their relationships with each other clearly. Especially in areas where low-velocity glacial drift is accumulated unevenly near the surface, the corrections can be difficult to make reliably. With a reflection flattened, the uncertainties and residual irregularities are eliminated. For instance, a reef may be more detectable when structure is removed by flattening. The structure may influence the situation, of course, and a structural section may also be needed.

A good reflection should be selected to be flattened. It can be at any depth on the section, and should be chosen primarily for quality. The only real purpose it has is to make easy and accurate the observation and timing of divergences between horizons that indicate whatever phenomenon is sought.

A quite different reason for flattening a horizon is to restore the section to some particular geologic time—after deposition of some for-

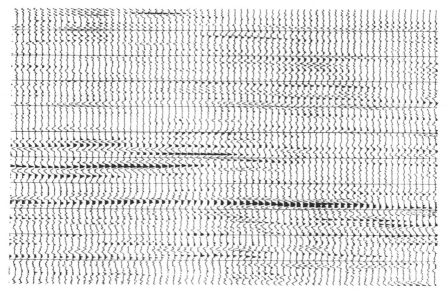

Fig. 7-23 True amplitude section Credit: Western Geophysical Co. of America

mation. This assumes that the formation was deposited flat. In this case, the selection of the horizon to be flattened is done on a geological basis. It should be the reflection from the formation desired, if that is pickable. If not, then the nearest reliable pick that appears to conform to the desired formation can be flattened. If it is difficult to flatten a horizon near the one you want flattened, then it may be helpful to first flatten whatever is the best pick on the section. Then, in another step, the picking and therefore flattening of the desired formation may be easier, with the smoother section to work on.

It may be useful to make several sections, with various horizons flat. They can then represent a sequence of geologic times, and can be used in a study of the geological history of the area (Fig. 7-24).

In 3-D work, there is another reason to flatten. A horizon can be flattened over a three dimensional area, and a time slice made at that level. With the horizon flat, the time slice is all on that horizon. Differences in amplitude or other characteristics of the reflection that can be seen on the slice may represent lithologic variations in the formation.

So depositional history, near-surface seismic problems, or lithology of a layer can be investigated by flattening. This is particularly easy with an interactive interpretation system.

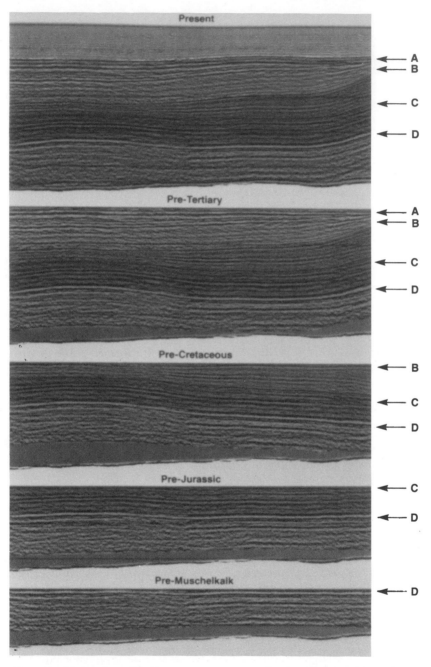

Fig. 7-24 Flattening of horizons   *Credit: PRAKLA-SEISMOS GMBH*

Dip Rejection

Dip rejection, also called a moveout filter or velocity filter, removes selected dips or apparent dips from the section. It has two main uses. First, in an area of rather flat bedding, steep noise lineups can be removed to clean up the section. Much of the random noise will appear to the computer as steep dip, as a wiggle on one trace will have some milliseconds of difference in time separating it from a nearby wiggle on the next trace. A person would look for a lineup across a number of traces to represent dip, but a simple program can have the computer look for "dip" just from one trace to the next. So if you know there aren't steep dips, the machine can throw out anything that looks steep. It will then throw out random noise and also diffractions (Figs. 7-25 and 7-26).

The second use is for taking out multiples in some geological situations by removing certain dips. For instance, multiples of flat layers in the shallow part of the section may obscure more steeply dipping reflections in the deeper part of the section. A section can be made

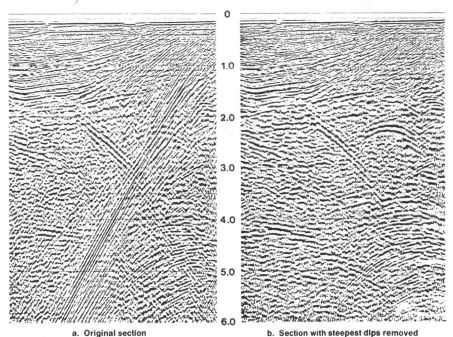

| a. Original section | b. Section with steepest dips removed |

Fig. 7-25a Original section   *Fig. 7-25b Section with steepest dips removed*

Fig. 7-25 Dip rejection to remove diffractions

Credit: Digicon Geophysical Corp.

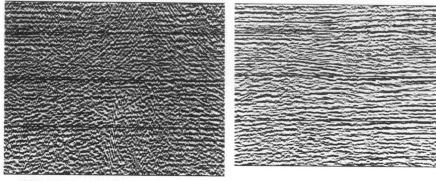

a. Without dip rejection   **b. With dip rejection**

Fig. 7-26 Dip rejection to remove noise   *Credit: Digicon Geophysical Corp.*

with the flatter dips eliminated. Then the flat multiples in the deep section are weakened, so the steeper dips can be seen. This weakens the shallow flat reflections, which are real, but a combination section can be made, showing flat dips in the upper part, and not in the lower.

In both cases you are assuming you know what the section should show. This is, of course, dangerous, as what you "know" may be incorrect. So dip rejection can sometimes be very helpful, but should be used with caution, and when used, viewed with some suspicion.

Velocity Determination

Velocity was discussed briefly in Chapter 2 and had a place in Chapter 5, Data Processing, in the special context of obtaining velocity information from seismic data, to enable CDP to stack properly. Now we need to look at it from the viewpoint of the interpreter, for converting seismic times to depth and for identifying horizons.

All velocity calculations use one simple basic formula—velocity times time equals distance, $VT = D$. That formula just says that going a certain speed for a certain length of time takes you a certain distance. Multiply time by velocity to get distance, or divide distance by time to get velocity, $V = D/T$. Or divide distance by velocity to get time, $T = D/V$.

Here is an example. Assume a velocity of 10,000 ft/sec, and a seismic reflection at 1.231 sec. How deep is the formation? O.K., multiply

10,000 by 1.231, and you get 12,310 ft. But the sound went down to the formation and also came back up, while you just want to know how far it is down to the formation. So you have to divide the answer by two, to get the depth. That's better, 6155 ft.

Now if 10,000 ft/sec is a velocity that applies to that reflection over a large area, you'll be multiplying a lot of reflection times by 10,000, and dividing by two. So to save time and reduce the number of errors, you might as well multiply by 5000, doing both operations in one step.

This using half the velocity is commonly used and sometimes leads geophysicists to talk about half-velocities. But although the whole velocity is not used much in calculations, it is almost always used in discussing velocities, to avoid confusion.

There was a reason for selecting 10,000 ft/sec as an example. That velocity is a good enough guess to use when you don't know the velocity, or when you just want a rough idea of the depth, or of relative depths. For that kind of overall depth, just multiply the time by 5. Notice that 5, rather than 5000. That works by using the time, not in seconds, but in milliseconds. Instead of multiplying 1.231 sec by 5000, you are multiplying 1231 ms by 5, and getting the same answer. If there is velocity information handy, a more accurate estimate can be made by using 4, or 6.2, or whatever, but in general 5 gives an idea of what is going on. It is saying that on the average, for every millisecond of two-way time, the horizon is 5 feet deeper. If there is a handy rule of thumb like that for an area, it pays to memorize it. In some areas a rule of velocity is that every millisecond of time adds one meter of depth.

On a section, the times have been corrected to an arbitrary datum plane which is noted on the labeling of the section, so the depth you get is depth below that datum. Offshore, the datum plane is usually sea level, or near it. Some sections are corrected to sea level, and others are left with the start of the section's timing at the average of source and geophone depth in the water. This is preferred by some, who do not like to tamper with the data any more than necessary. It makes only a small difference in absolute terms, something like 30 or 40 feet, and is uniform within one shooting program, so it can often be ignored.

Some approximate velocity information can be used in well-known areas. If a reflection can be recognized by familiarity with the general area, then a line shot through a well's location can be tied to the well. That is, the reflection may have been identified in the past at other wells. If it has a distinctive shape, then it may be recognized from experience in the area as being the reflection from a certain formation, that is, a reflection that had been identified elsewhere in the area. The time of the reflection on the section and the depth of that formation in the well give the velocity to the formation at that location, even though there has not been a velocity survey in the well. Just obtain the depth of the formation in the well below the datum used on the section, and divide by the time, to get the half-velocity. If there are lines shot at other wells in the vicinity, lateral changes of velocity can be determined.

Velocities from NMO data are also useful. But, especially for deep horizons, they may not be accurate enough for conversion of time to depth. They were described under "Velocity Analysis" in Chapter 5, Data Processing. More precise and reliable velocity information comes from a sonic log combined with check shots.

Velocity Survey

Check shots in a well are made by firing a seismic source near the surface, and detecting it with a geophone suspended at a series of depths in the well. This gives an exact measurement of the times required for the sound to go to known depths. They are called check shots because they are shots to check a sonic log.

A special check shot crew is arranged for, and also a well-logging unit to provide a cable that can be used to lower the geophone. The survey is performed as soon as the well is released for it, so as not to run up rig standby charges. If the well is offshore, an air gun is lowered into the water from the rig floor. On land, usually a marine-type air gun is placed in the mud pit of the well. A geophone, built into a streamlined unit, is attached to the logging cable and lowered to the bottom of the hole. Then it is pulled up in stages, preferably evenly spaced, say every 500 feet. A spring presses it against one side of the hole, so the geophone has good contact with the earth. At each stop the air gun is fired, and the time taken by the sound to reach the downhole geophone is recorded. A recording is made of each shot.

The shots are not right at the well, so the sound travels a slanting path, and the times must be corrected to vertical paths by trigonometry. However, for deviated wells, the shots can sometimes be placed directly above the geophone at the various depths recorded. The paths are then vertical. For deviated wells, the geophone may be gimbal-mounted within the probe, to keep it upright.

At each recording position in the well, the seismic time to that depth is measured, and the velocity is calculated. This is done by reading the first breaks, the first energy to reach the geophone at the different positions of the geophone in the well. The times of the first breaks may be plotted on a T-D chart, a time-depth chart, with depth plotted vertically and time horizontally (Fig. 7-27). These one-way times are doubled to be compared with a seismic section.

This provides good velocity and identification information for a seismic line run through the well location. With this information, reflections on a section can be identified just by their times. On a zero-phase section, a reflection appears strong at the time of the formation—twice the time calculated from the check shots. On a minimum-phase section, a reflection will appear strongest at a later time. Something like 30 to 60 ms is required for the energy to build up to a good peak or trough on the section.

The check shots give data only at isolated points in the well. To make the data continuous, first break times from the shots are combined with a sonic log. The sonic log is continuous, but does not have the overall correctness of the check shots. A velocity survey made by combining the two has both qualities.

A sonic log is used by geologists and log analysts for porosity information, but the quantity it actually measures is the velocity of sound in the near vicinity of the well. Porous rocks transmit sound more slowly than solid rocks, because porous rock consists of a combination of high-velocity rock and low-velocity fluid in the pores. So the velocities correlate well with porosity.

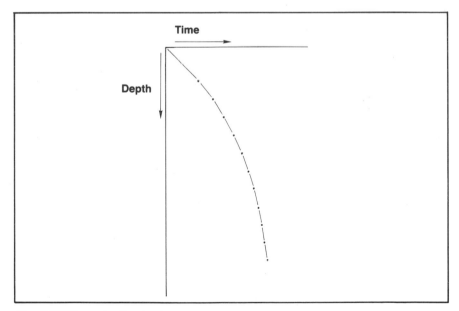

Fig. 7-27 Time-depth chart

A sonic log is made with an instrument containing, in its simplest form, a sound source, or beeper, in one end, and a geophone in the other. It is lowered by cable to the bottom of the well. The source beeps, and the first of the sound from it to arrive at the geophone is the sound that went through the rock alongside the well bore. It is refracted at a critical angle and arrives before the sound that goes directly through the slow-velocity drilling mud in the well. The instrument is slowly and smoothly pulled up the hole, with the beeper emitting signals frequently. The times recorded at the different depths are plotted as a continuous log in seconds/foot, the reciprocal of velocity.

In its more standard form, the instrument has one sound source and two receivers. The instrument is actually a very small-scale refraction spread. Also the instrument itself is designed to transmit at a slow velocity, by the use of rubber segments, so signal traveling through its metal body won't be recorded first. Subtracting the time for sound to reach the near receiver from the time to the far receiver leaves a segment of the path through the rock.

There is a cumulative error, though, in the overall velocity as assembled from these recordings. Mud invasion of the rock near the hole, and the washing out of the hole (enlargement of parts of the hole during drilling) may modify it. To correct for these inaccuracies, the velocity survey, with its velocities measured by firing shots at the surface, can be used for the velocities overall, and the sonic log corrected to it. The log then becomes a CVL, continuous velocity log, calibrated sonic log, or sonic log with reference marks. Then the shooting is referred to as check shots, to check the sonic log.

In this combination is the reason for firing the shots with the geophone at arbitrary intervals in the well. At one time, before sonic logs were developed, the shots provided the only data for the survey. So they needed to be taken at formation tops, to yield velocities of formations, rather than lumping parts of formations together. Now, with the shots used only to adjust the sonic log data, they are evenly spaced and the sonic log gives the velocity changes at formation contacts.

Synthetic Seismogram

When a velocity survey is made in a well, very detailed velocity information becomes available. Velocities are obtained from the sonic log, calibration of the velocities from check shots, and densities from a density log. A density log is made by lowering in a well an instrument that has a radioactive source of gamma rays, and a detector that measures the gamma radiation that is returned from the rock near the well. It differs from a gamma-ray log, that uses a detector but no source, measuring the radioactivity in the earth.

As seismic reflections are a result of velocity and density contrasts, there is enough data to calculate just where in the section there would be seismic reflections, and their amplitudes and polarities. So wavelets can be put together to make a synthetic seismogram, which is a theoretical seismic trace, like one that would be expected on a seismic section at that location. Being made from the well data, the synthetic seismogram is exactly fitted to the well. The word "seismogram" is an outdated and otherwise rarely used word in oil exploration, a word that happened to be preserved in this one context.

To make a synthetic seismogram, the velocity and density data, or the velocity data alone if density information is not available (Fig. 7-28), are used to calculate a reflection coefficient at each change in velocity

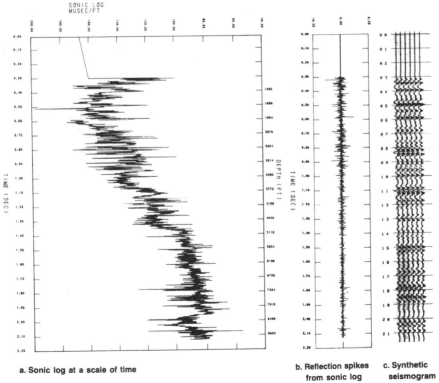

a. Sonic log at a scale of time

b. Reflection spikes from sonic log

c. Synthetic seismogram

Fig. 7-28 Steps in making synthetic seismogram
Credit: Western Geophysical Co. of America

and density. The reflection coefficient of an interface is a mathematical expression of the kind of reflection that will be produced at that interface—amplitude, that is, how far the trace swings; polarity, which way it swings. The polarity of a minimum-phase reflection on a section is a matter of which way it swings first, for a zero-phase reflection it is which way its strong central swing goes.

A wavelet is selected that is thought likely to match the basic wavelet of nearby seismic data. For an interface the computer program calculates a wavelet with the proper maximum amplitude and the proper polarity. Then it does the same for the next interface. But the second interface may be only a few milliseconds from the first, so the two wavelets overlap. Never mind, the same thing happens to the wavelets in the ground. The two are convolved, combined into one trace, as was described in "Convolution", in Chapter 5, Data Processing. This is continued for the rest of the well survey, combining wavelets of varied amplitudes and polarities to make a seismic trace.

Then, using the reflection coefficients for energy traveling upward, the part of the sound that is reflected back down to produce multiples can be calculated. So multiples can be added to the trace.

The trace is then displayed at the same vertical scale as a section, so it can be compared directly with sections. Four or so copies of the trace are plotted side by side to look a little more like a section, rather than a single trace.

Now suppose there is a synthetic seismogram from a well, and a section that was shot across that same well. Comparing the synthetic with the section can do several useful things, if the section and synthetic look alike. Sometimes they don't. Assuming the two are similar, reflections can be reliably identified. The synthetic is made at the time scale of the section, but depths can also be plotted on the synthetic. Formation tops from the well can be plotted on the depth scale by a geologist familiar with the area. The reflection on the synthetic can be expected to be strong at the interface that produced it, or at a lag of about 30 to 60 ms later, if the synthetic is minimum phase. And this reflection should look like a reflection on the section at that point. If the match isn't at that point, the synthetic can be slid up and down on the section, within the possible lag, to find the best point of overall similarity of reflections between the two.

The reflections and multiples were calculated separately for the synthetic, so they can be displayed separately. It is useful, on a synthetic seismogram, to display a few traces each of primary reflections only, multiples only, and the combination of the two. The combination or the primaries only will normally look most like the section, depending on how well multiples have been attenuated on the section. The primary-only and multiple-only groups of traces will make it easier to distinguish between primaries and multiples on the section.

The displays can also include another set of traces in those three forms, but with opposite polarity. And they can all be displayed at different filter settings. Then a person comparing the synthetic with a section can try various displays to see which looks most like the section.

Displays of synthetic seismograms by different companies take different forms, and are made in differing degrees of sophistication. They can range all the way from using sonic data only, no density; and calculating reflections only, no multiples; to the most complete. They may be at any common scale of seismic sections. So they cannot all be expected to look alike, or to be equally useful. It may help, before ordering a synthetic seismogram, to discuss scales, purposes, and special requirements with the company that will make it.

Synthetic seismograms sound like just the thing to establish definite identifications of seismic reflections. But there is a flaw. They often do not look much like the sections right at the wells. However, advances in making synthetics continue to make them better. One problem is the condition of the hole in which the sonic log is taken. There may be zones where soft rock washed out, leaving the hole larger than normal, so the reading isn't correct. Careful editing of the log will help. It is edited by comparison with a caliper log that shows hole diameters, and with other logs, of types that are not so strongly affected by washouts.

Synthetic Sonic Log

As described in the preceding segment, a synthetic seismogram is a seismic-like trace produced from the velocity information in a sonic log (and maybe data from a density log). Comparing it with seismic traces partly bridges the gap between well data and seismic data.

For another comparison between the two, something like a sonic log can be made from a seismic trace. But a sonic log is a collection of velocities, so making a synthetic sonic log from a trace requires much better velocity information than is usually obtained in seismic work. To get this velocity information, something beyond the resolving power of velocities obtained from normal moveout is needed.

The amplitude of a seismic reflection is dependent on the velocity contrast of the reflecting interface. If the velocities of the two layers are known, the strength of the reflection can be calculated, as is done in making a synthetic seismogram. Then, turning that around, if the amplitude of the reflection is known, maybe the velocities can be obtained. At least, if the amplitude and one velocity are known, the other velocity can be calculated. One velocity can be obtained by starting at the top of the record, where a near-surface velocity can be obtained fairly readily. Then use that velocity and the amplitude of a reflection at the base of that layer to obtain the other velocity at the interface. Having the second velocity, a third, at the next reflecting surface, can be determined, etc.

For so much to depend on the amplitudes of reflections, those amplitudes must be accurately measured. This requires recording by floating point or binary gain, methods that determine the true amplitude, not adjusted to make reflections strong all down the record. Also, for the measurements to show velocity changes in fine detail, the recording should be broad band, that is, with more frequencies recorded than are normally used for seismic data. In particular, higher frequencies must be included (Fig.7-29). Normal moveout velocity data

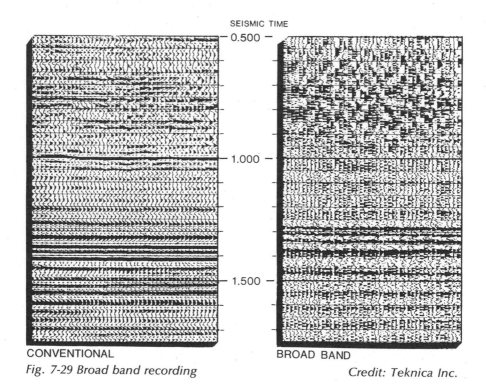

SEISMIC TIME

0.500

1.000

1.500

CONVENTIONAL   BROAD BAND

Fig. 7-29 Broad band recording   *Credit: Teknica Inc.*

can also be added to the information obtained from reflection amplitudes, to improve the overall correctness of the velocities.

A section can then be made up of traces that look like a series of sonic logs, but considerably smoothed. Even broad-band seismic data does not have nearly the fine detail of a well log (Fig. 7-30).

As an additional aid to interpretation, the velocities along the traces can be contoured by computer. Velocities tend to be in long narrow lenses on a section, because they apply to layers of rock, and change occasionally as the layer changes depth or lithology. In particular, sands, porous zones within reefs, etc., can be interpreted on a synthetic sonic log section. And if the line crosses a well, the sonic log from the well can be smoothed to match the smoothness of the traces, converted to fit the time scale of the section, and superimposed on the section for calibration of the seismic-derived synthetic sonic log traces.

As reflections are generally poor in the deepest part of the section, resolution of small features by synthetic sonic logs gets poorer with depth. The technique is best at shallow depths, but resolution of strong events below 10,000 ft. is common.

VSP

A drilled well offers an opportunity to make a seismic section different from the normal type—a section with the energy source on the surface as usual, but with the geophone positions down the hole.

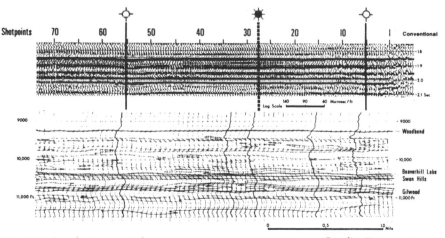

Fig. 7-30 Synthetic sonic logs   *Credit: Teknica Inc.*

Check shots for a velocity survey are made in this way, but only first breaks are recorded—not enough data to form a section. A VSP, vertical seismic profile, is made like a series of check shots, with a geophone lowered into the hole and shots initiated at the surface as it is pulled up to different depths in the hole. But in a VSP the source is farther from the well, more data is recorded than just the first breaks, and more shots are fired so there are more traces.

A VSP, like a section recorded from a surface spread, has both primaries and multiples. But it differs from a normal section in the directions the energy is going when it hits the geophone positions. A spread of geophones on the surface receives energy that is going upward from the subsurface. But a VSP is recorded at different geophone depths in the ground. Sound can reach those geophone positions on either the way up or the way down. This creates a major distinction in the types of lineups of energy on a VSP. Some are from upgoing energy (upgoing when it arrives at the geophone), and some are from downgoing energy. Alignments from those two types of energy slant in opposite directions across the VSP. Corrections can be made to enhance either direction and weaken the other. The main usefulness of a VSP depends on the two types of events and the corrections applied to them.

We'll take up several kinds of downgoing energy first, then several kinds of upgoing energy, then the corrections and the uses of VSPs.

The first breaks are the first energy that arrived at each depth, the downgoing energy that went in a direct line from the shot. In check shots, only first breaks are used at the different depths, and the times of the breaks may be plotted with depth shown down the vertical axis and time horizontally, as a time-depth chart. The curve starts out at a downward slant, and becomes more nearly vertical with depth. Moving down the scale to a depth and across to the curve, the time for that depth can be read. The first breaks in a VSP are displayed in the same arrangement as the first break times are in a time-depth chart. The time for a depth can be read in the same way.

On a normal section, the traces hang down vertically, but on a VSP, the traces can be extended horizontally from the first breaks, letting the depth scale be vertical. The display consists of the curving line of first breaks, with horizontal traces sticking out from them (Fig. 7-31).

On the VSP plot, other alignments of energy can be seen parallel to the first breaks. What are they? They are multiples that were reflected up and down again on the way (Fig. 7-32)—think of them as second breaks, third, etc. Some of the later ones even went beyond the depth of the well and bounced back up before starting down again (Fig. 7-33). But all of them are downgoing energy. They were on the way down

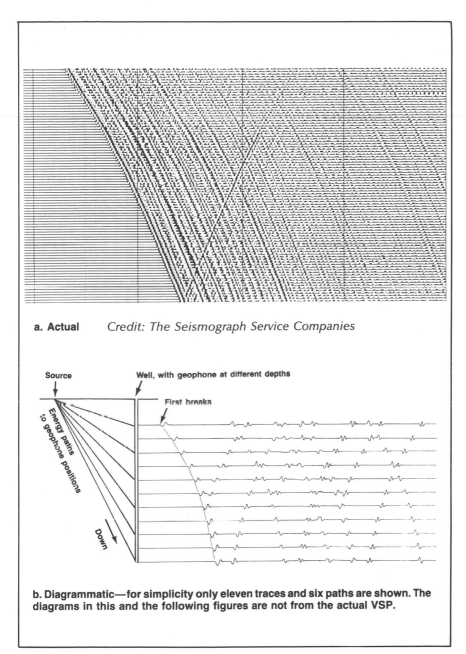

a. Actual   *Credit: The Seismograph Service Companies*

b. Diagrammatic—for simplicity only eleven traces and six paths are shown. The diagrams in this and the following figures are not from the actual VSP.

Fig. 7-31 Uncorrected VSP

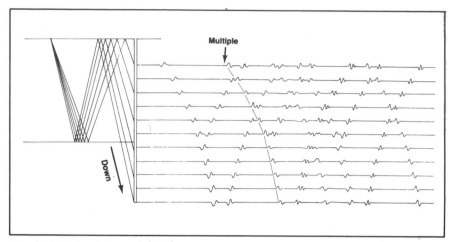

Fig. 7-32 Downgoing multiple

when they arrived at the geophone, even though some wandered below it on the way. *Downgoing energy is from first breaks or multiples of them.*

There are other lineups on the display. Instead of being parallel to the first breaks, they slant down to meet them, crossing the parallel ones on the way. They are upgoing reflections that arrive at the geophone from below (Fig. 7-34). Think about one of these lineups, at the point where it meets the first breaks. At that point, the geophone is at the level of a reflector, and the first break is at that level. At the next

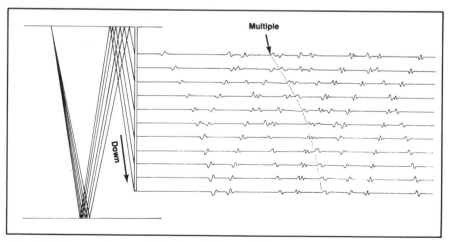

Fig. 7-33 Downgoing multiple from below well

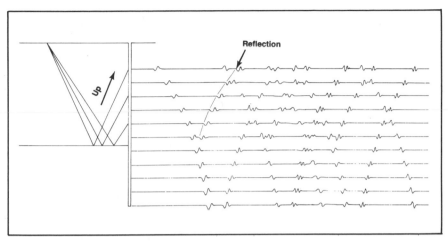

Fig. 7-34 Upgoing reflection

geophone position up the hole, energy from that horizon will arrive at the geophone by being reflected from a point a little distance away from the well. As the geophone moves up the hole, the reflection point moves farther from the hole, toward the source position. If the geophone was raised to the top of the ground, the reflection point would be midway between source and geophone. It isn't raised that high, so the reflection is that short segment from the well to some point short of the midpoint between the source and the well. The shallower reflections are shorter segments, as they meet the first breaks sooner than the deeper ones do.

This upgoing energy can also go up to the geophone positions from below the bottom of the well (Fig. 7-35), and from multiples (Fig. 7-36). *Upgoing energy consists of primary reflections and multiples of them.*

For a specific reflection, the deeper the geophone, the shorter the distance energy travels upward, while the way down stays the same. Like correcting for uphole time, the two paths need to be made equal. The way uphole time is corrected (as in the chapter on data processing) is to subtract its equivalent from the reflection time, making the two paths equal. The paths in VSP data are also made equal, but by adding the difference to bring the reflection path up to the surface.

Adding the first break times to the section changes the first break curve, giving it double the time difference from top to bottom. Also, by making the two ends of the path equal, it brings the upgoing energy into line, so that, if the reflecting layer is flat, its reflection looks flat.

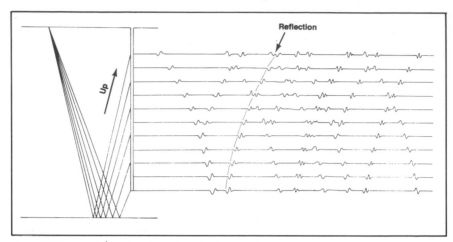

Fig. 7-35 Upgoing reflection from below well

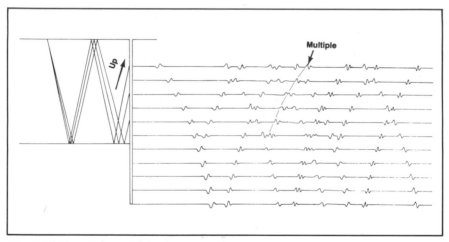

Fig. 7-36 Upgoing multiple

With this correction made, it is convenient to turn the display so the time axis is vertical (Fig. 7-37). Then it looks like, and can be matched with, a normal, surface-recorded section. It is also in two-way time, like a normal section. The horizontal events on it, like those on a normal section, all arrived at the geophone by reflection from below.

With the reflections fairly flat, they can be enhanced by applying a dip rejection process that weakens the steeper events. Then the reflections from below are easier to see, and the multiples of the first breaks, from above, are unnoticeable.

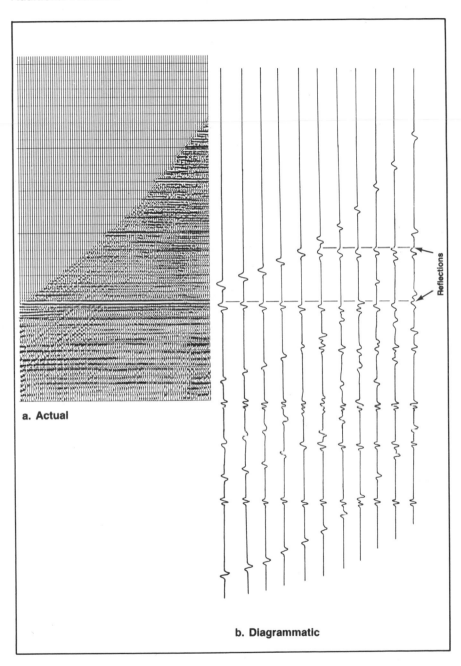

a. Actual

b. Diagrammatic

Fig. 7-37 Doubled first breaks   Credit: The Seismograph Service Companies

For any reflections on the VSP, there can also be multiples of those reflections. The geometry of VSP includes a way of distinguishing some multiples from primaries. A reflection that is shallow, and therefore short, because it is cut off by the first break curve, will have a multiple that is equally short, although the multiple appears at a later time on the VSP (Fig. 7-38). It took longer in its extra bounce, but it arrived at the same geophone location. So a reflection that doesn't reach the first breaks at its depth, but is only as long as some shallower reflection, is likely to be a multiple of that reflection. In addition to being a means of identifying multiples, the VSP shows, beyond the short length of the multiple, what the section would look like without the multiple. This multiple detection works only for primaries as far down as the deepest geophone position. Below that, the VSP is more like a section shot on the surface.

A VSP can be used to identify events on a seismic section. Any depth in the well can be located on the first-break curve, and a reflection at that point on the VSP can be matched with one at the same reflection time on the section. Multiples identified as such on the VSP can also be matched to their equivalents on the section. And, of course, a VSP has all the information a set of check shots has, but in more detail, so it is similarly useful as a source of velocity information.

Another use is in "looking ahead of the drill". There are deep reflections on the VSP that are from formations below the bottom of the well. They have better detail than a seismic section has, both because of trace spacing and because the geophone was nearer to the formations for the part of the trace close to the well. This advantage is sometimes used during the drilling of a well. The well is drilled to some depth, then a VSP is run in it and interpreted, then the well is deepened, deviated, abandoned—whatever is called for by that additional information.

The fine detail of the VSP and the fact that much of it is made with the geophone close to the formations allow it to be used to investigate the effects on a normal section of weakening of energy or distortion of wavelets in passing through the subsurface.

All of that was done with the VSP corrected by doubling the first break times. There is another useful correction that can be made. It is subtracting the first break times. As in an uphole time correction, it makes the paths down and up equal, by cutting off some of the downward path. It puts the first breaks all in a horizontal line, at the top of the section (Fig. 7-39). And in so doing, it makes their multiples, the multiples arriving at the geophone from above, horizontal also. They were originally parallel with the first breaks, so they become horizontal when the first breaks do. As with the reflection display, this one can

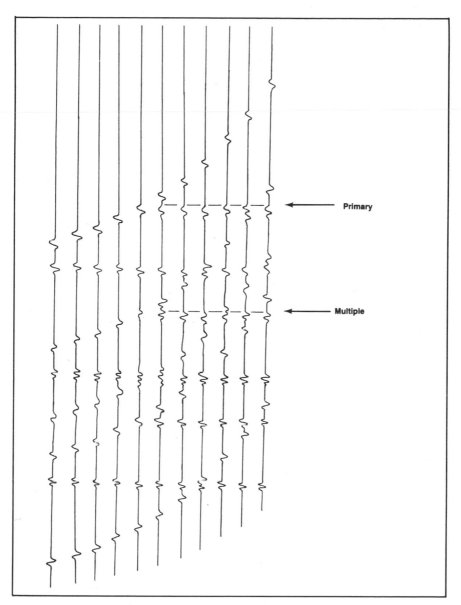

Fig. 7-38 Short multiple

have the horizontal energy enhanced and the steeper energy weakened. It is then a display of multiples only, and can be used to detect multiples on a normal section, or on the display of the VSP that shows reflections. In either case, it is placed against the section to be inves-

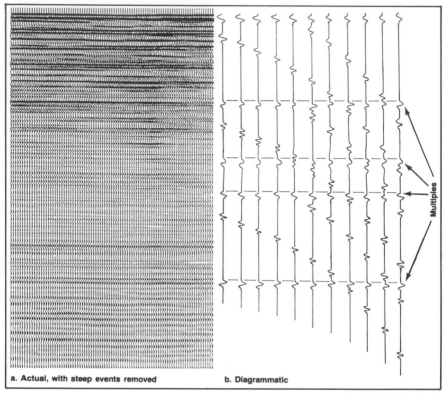

a. Actual, with steep events removed   b. Diagrammatic

Fig. 7-39 Subtracted first breaks
        Credit: The Seismograph Service Companies (a. Actual)

tigated (Fig. 7-40). It is slid down so its top is at a reflection on the other section. Then its multiples show where multiples of that reflection can be expected.

A VSP is a way of arranging sources and receivers, just as shooting lines on the surface is. So, wherever an open well is available, a similar variety of applications can be employed. The source of energy can be placed in different directions from the hole, to record energy in those directions. And the farther it is from the well, the longer the extent of subsurface the segments of reflection will cover. The closer it is, the nearer together the reflection points, like closer group spacing on a line shot on the surface. Before a VSP is shot, a distance for the source is selected, balancing these two factors to fit the particular problem. The source location is usually chosen for closer spacing than would be used on a seismic line. This gives good detail near the well. If a well is deviated, then, as with check shots, sources can be placed straight

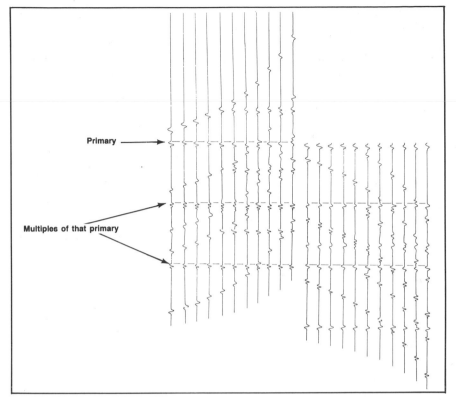

Fig. 7-40 Multiple identification

above the geophone at each of its positions in the hole. Or shots can be in a line to or through the well location, or over an area around it. A synthetic VSP can be made to check the interpretation of a VSP.

Modeling

Modeling is a term used in seismic exploration for producing artificial data to compare with actual data obtained by shooting, or to see what data might be obtained in a certain situation. It can include artificial seismic sections, synthetic seismograms, and synthetic VSPs.

One type of artificial seismic section is made from synthetic seismograms placed side by side to form the section. This can be used to determine the point at which a pinchout of a formation occurs. An apparent pinchout on a section is not at the point where a formation thins down to zero thickness and ends. Rather, at some place before that, the reflections of the top and bottom of the layer get so close together that their wavelets blend, and the two reflections are no longer separate. The combined reflection continues, but its character changes as the wavelets, getting still closer together, blend in different ways. But that doesn't show where the formation pinches out.

Suppose there is a well that encountered 433 feet of a formation, and the formation is thought to pinch out not far away. It might be useful to find out exactly where the formation ends. Also suppose that a synthetic seismogram has been made from a sonic log (or sonic and density logs) in the well. The log shows 433 feet of the formation. Part of the indication of that formation can be cut out of the log with scissors, and the remaining pieces taped together. Then a synthetic can be made from the spliced log. This is the synthetic that would be expected in a well in which that formation was a little thinner. Then some more of the formation can be cut out of the log, and another synthetic made. The process can be continued until all of the formation has been removed. The series of synthetics can be displayed side by side in order of diminishing thickness of the formation. This section shows what appearance is to be expected where the formation is 200 ft. thick, 100 feet thick, 10 feet thick, etc., and where it is absent. If no other factors interfere, then this section made of synthetics can be used to determine the various reflection characters as the formation thins.

Similar treatment can be used for reef thickness or salt thickness. A log that encountered an edge of a reef or salt swell can be modified by increasing that thickness to a possible thickest part of the feature, and then thinning it down the other side. Synthetic seismograms can be made of the modified logs to give a realistic picture of the feature in the form of a seismic section. In another modification of sonic logs, part of a log can be cut and shifted to one side or the other to represent a higher or lower velocity.

The actual cutting with scissors is not necessary, but was used to make the point clearly. The modifying of the logs is done on computers by data processors.

By a similar process, logs can be modified and synthetic seismograms made from them so as to show a particular formation dipping when they are placed side by side. Or the dips of several horizons can be given different angles. The artificial dips can be useful if the artificial section is to be matched to a section that has considerable dip on it.

Another type of artificial section is made by ray tracing, that is, calculating the paths of seismic energy. If the velocity of sound in a rock layer is constant, the path of energy through the layer is a straight line. If the velocity increases steadily with depth, the path is a predictable curve. When the sound enters another layer, its angle changes by refraction in a predictable way. So if the thicknesses and velocities of the layers are known, the routes of seismic energy through them can be determined.

When the geology is not known except for a seismic interpretation, the model can be made from the interpretation by ray-tracing. The interpretation must be rather complete, with the reflection times converted to depths and with velocities estimated for all the layers. Then the paths of reflections that would be recorded from that geological situation are calculated (Fig. 7-41). Then those reflections are plotted as an artificial section. That section is compared with the original section. If they are alike, then the interpretation was probably made fairly correctly. If they differ in important features, then probably the interpretation should be revised. The revised interpretation can then be modeled, and the new model compared with the original section. This process can be repeated to produce models more and more like the original.

Most models use raypaths based on the source and receiver being at the same point, like a NMO-corrected section. Or a model can be made to include all the source-to-geophone offsets of CDP shooting.

Modeling is useful in bright spot analysis. The velocity effect of gas in a formation can be included in the calculations, to see if that aspect of the original section is duplicated in the model, and to determine from it an idea of the percentage of gas in the rock.

Modeling to simulate a 2-D section may necessarily be done in two dimensions. With no other information than the section provides, that is the only kind of ray tracing that can be done. But an interpreted map

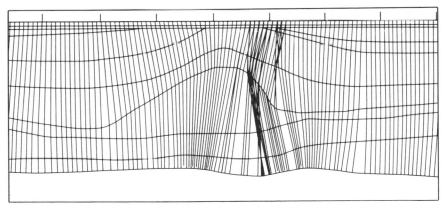

Fig. 7-41 Ray tracing   *Credit: Western Geophysical Co. of America*

made from the lines in an area contains structural information to the sides of the single line, so a 3-D model can be made of the line. This includes the raypaths of sideswipe, out of the vertical plane of the section. And, of course, a 3-D model can be made from an interpretation of 3-D shooting.

The modeling process is costly and time consuming, so it is normally not used on a regular basis for checking interpretation, but saved for critical points. Determining and/or assuming all the velocity data is a lot more interpreting than is necessary in normal work, where the configuration of the beds is the main thing that is needed.

Another use of modeling is in determining efficient ways to shoot 3-D surveys on land. The survey is so expensive and must be so thoroughly planned that it is worthwhile to make a considerable effort beforehand to improve the planning. Information about dips and other conditions can be obtained from 2-D lines and other sources, and put into a computer. Then modeling can simulate shooting the area with various line spacings, group spacings, etc., to determine the best way to do the real shooting of the area. Offshore, the preplanning for 3-D isn't so necessary. Cable lengths and group spacings are more or less fixed, and any gaps in control caused by currents or mis-tracking can readily be filled in by shooting more lines.

CHAPTER 8

Interpretation Procedure

Seismic interpretation is the process of determining information about the subsurface of the earth from seismic data. It may determine general information about an area, locate prospects for drilling exploratory wells, or guide development of an already-discovered field. This chapter will describe the identification, picking, timing, and contouring in a normal interpretation, as done on paper and on a video screen.

People have looked over my shoulder and asked, "How do you read all those lines?" Thought of that way, it's daunting. There are a lot of lines on a section. How can a person understand all that? The answer is that it isn't necessary to "read" everything on a section. Instead, certain features of the section are selected to interpret.

An interpreter concentrates on certain lineups of wiggles or color crossing the section. It is necessary to somehow decide what formations some of them represent, how they go higher and lower in crossing the section, where the formations are broken—things like that.

The most common activity in interpretation is picking a horizon. A horizon is a reflection that appears on sections over some geographical extent. The reflection may be identified as representing a certain geological formation, or only as being a reliable reflection. The reflection is picked, that is, marked at shot points along the vertical sections over the area. The picks are timed, by reading the reflection times. These times can then be plotted on a map and contoured, to show locally high places or other features that may be prospects for drilling. Or the contours may be traced from reflections on time slices.

There are a number of special problems in interpretation—identifying the reflections, staying on the same reflection throughout an area, presenting the results, etc. The approach used in this book will be to describe how these steps are taken with paper sections, and then to describe the faster, easier ways they can be handled interactively with a computer.

Identification

When a seismic line is shot, if record quality is good, there are a number of reflections on the resulting section. Similarly, when a well is drilled, a number of formations are encountered. Now if the line and the well are near each other, at least some of the reflections are from some of those formations. But how do you know which is from which? The reflection times are measured in seconds and the depths to the formations in feet or meters.

If you had an absolutely perfect knowledge of the velocity of sound at that location, you could calculate the relationship and determine exactly how deep a reflection is, or how much time is required for sound to go down to a formation and return. Then you could relate a certain reflection with the formation it came from.

The nearest approach to this perfect velocity knowledge occurs when a line is shot across the well location, and a velocity survey or VSP is made in the well. Then, multiplying half the reflection time by velocity gives the depth to the reflector. Or dividing the depth by the velocity, and doubling, tells what reflection time represents that formation top.

If there is a velocity survey in one well and a seismic line does not go through the well, but does cross another well in the area, then a correlation from one well to the other can provide a sort of velocity information at the second well, and it can be related to seismic reflections.

One of the most common ways of identifying reflections is to compare the section with another section on which identifications have been made. This becomes poorer as the two lines get farther apart.

Synthetic seismograms are frequently used to identify reflections. Made from well logs, they have exact formation times, and can be correlated directly with seismic sections.

A positive correlation between a seismic section and a well can be made with a VSP. One edge of the VSP is in depth to match the well, and another is in time so it can be correlated with the section.

Velocities obtained from NMO are sometimes good enough to identify reflections in a general way. In some areas that have not been explored much, this may be the only means of identification available.

In some remote areas with no wells, projecting surface geologic data downward may be the only way to relate seismic reflections to formations.

In general, a geophysicist can not put a finger on a reflection and say, "There is the So-and-So Formation". But this eyeball recognition can work in familiar areas, where much experience has been gained, and a form of it operates in areas known to have only one or two good reflectors. If just two pickable reflections are expected in an area, and two are visible on a section, it's easy to decide which is which.

Continuity

Once a horizon has been selected for picking, whether identified or not, there must be some means of picking the same horizon throughout the area, or at least over some part of the area.

If the record quality is good, and there aren't any complicated situations along the line, a colored pencil mark can be drawn on the reflection from one end of the section to the other.

A comment on the pencils. The especially erasable colored pencils allow the interpretation to be made in distinctive colors for different horizons, and changed as often as needed. The colored pencils with built-in erasers on the ends are the erasable ones. But those erasers aren't good for use on sections—they're too abrasive. Soft erasers, either plastic or gum, are better.

The interpretation problem comes in when the record quality is not good, or when there is faulting or some other complicating factor. There is a starting place—the well location or other point at which the reflection was identified. As long as it is apparent that the same reflection is continuing without a break, it can be picked. Then if the reflection becomes poor on part of the line, making a break in continuity, a search can be made above and below it for better reflections. If there are some, the poor one probably conforms with them, so it can be drawn parallel to them. Then it may be found again farther along on the section, on a reflection that the drawn line meets or nearly meets (Fig. 8-1).

Also, the part of the reflection already picked can be extended with a straightedge, continuing the same dip, and reflections on the other side of the poor zone extended back toward it. A good meet of these lines extended from two directions indicates that they are probably on the same horizon (Fig 8-2).

Position on the section is useful in determining which is the same reflection after a gap. Naturally, if the uppermost strong reflection is being picked on the section, then after the gap the uppermost strong reflection is a candidate for being the same one. And if the horizons are

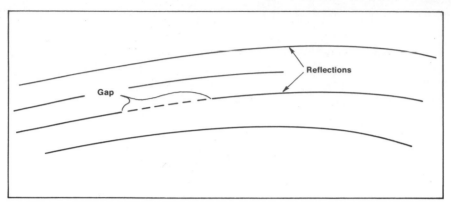

Fig. 8-1 Parallel other reflections

not dipping much in the vicinity, then one at the about same time or depth is likely to be the same one.

Occasionally an area will be encountered in which no continuous reflections can be picked at all. This may be because record quality is poor, or because the subsurface itself contains no continuous layers, but just lenses of sand and shale, as in parts of the Gulf of Mexico. In either of these cases, a line can be drawn on the section paralleling the discontinuous bits of nearby reflections. This line is a phantom horizon (Fig. 8-3).

Character

Then there is character. Reflection character is the shape of the wiggles that make up a reflection. A seismic trace is made up of just sine-wave-type wiggles back and forth across a central position, but the

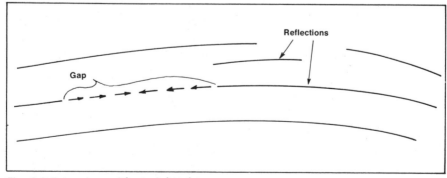

Fig. 8-2 Extension with straightedge

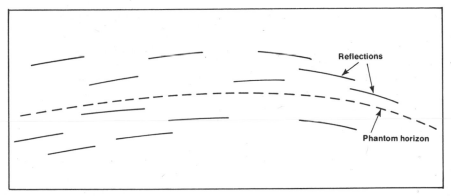

Fig. 8-3 Phantom horizon

wiggles can vary in size, so there is a good deal of variety in the shape that a band of energy, several peaks and troughs in sequence, can take. A reflection can be recognized by its having one large peak followed by a small one, two large and one small, or something (Fig. 8-4). A broad peak with a dip in its top is called a saddle. Two peaks or two troughs together are a doublet, or doubleton.

Almost any reflection is some combination of two or three of these features that persist for some distance on a section. The distinctive pattern of the combination is the character of the reflection. A charac-

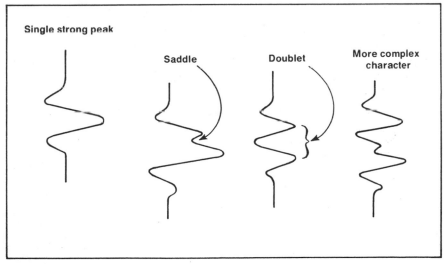

Fig. 8-4 Various reflection characters

ter often persists over many miles. An example is the two-peaks character of the Wabamun, recognizable all over the Rainbow-Zama area in
Alberta, Canada. In other areas, a recognizable character may carry for
only a short distance.

Character permits correlation of traces, just as well logs are correlated, so reflections with good character can be followed across faults,
poor records, gaps in control. The character on seismic sections is not
as definitely recognizable as are features on well logs, and not as
detailed and measurable down to the foot, either.

In most interpretation of sections, except where there are no problems, all of these techniques are used to some extent. In combination,
and used by an experienced interpreter, they constitute an effective
means of determining horizons.

Loops

Two-dimensional seismic lines are usually arranged in loops. That
is, they cross each other so a reflection can be followed from one line
to another, and on around to its starting point. On a regular grid of
lines, the loops will be rectangles. With other programs, they may be
irregular shapes with different numbers of sides. The idea of a loop is
(as in land surveying with an optical instrument) to check to see if the
information is still on the same level when it gets back to the starting
point.

Loop tieing is a good check on picking. If a loop ties, it doesn't
necessarily mean it is picked correctly—two errors could cancel each
other. But if a loop doesn't tie, the picking has to be wrong somewhere
on the loop.

When a loop doesn't tie, it may be possible to correct it by going
back to a point where a difficult decision was made, and trying an alter-
native choice. Another way is to check around the loop for just plain
errors. On occasion, there is nothing better that can be done than to
find the poorest part of the loop, assume that the error is there, and
change the interpretation there, to make the loop tie. That isn't good,
but it's worse to leave it mis-tied.

In a large area in the Gulf of Mexico, a grid of 2 by 3 miles or something like that was being shot. Two factors made good ties difficult.
There were a lot of faults in the area, but they did not show up seismically. And the subsurface tended to be made up of sands and shales
that were not in continuous layers. A careful interpretation was being
made, but it was slow, having to wait for accurate survey data, etc. We
needed some idea of the structure quickly, while the shooting was

going on, so I started a "sloppy map". To make it, I interpreted sections, mis-tied loops by 100, 200, 300 feet, then just adjusted the data to make the loops tie. The adjustment was made with no consideration for quality of data—just take 100 feet of dip out of one line, add 250 feet to another line, because it will make the loops tie.

The map wasn't technically correct, but it was continually ready as needed. And, after the more careful mapping was completed, the similarity between the two maps was pretty good. Certainly the salt domes were in the same places. The point of this story is that in some cases loop tieing alone can produce usable data.

For some situations, loops aren't necessary. A line shot from one well to another, to tie the well data, doesn't usually need the additional check of loop tieing. A long line, shot only for regional information, doesn't need to be so correct. And sometimes only an indication of a higher location near a well is needed.

If several lines are shot in the direction of dip, as in the steeply dipping areas along mountain fronts, they may be left separate, or may be tied with only one line joining them. If there are faults with large throw, the faults may make ties unreliable, so the loops don't produce any increase in confidence in the interpretation.

Formations and Reflections

Seismic sections are interpreted in terms of geological formations, and the reflections are identified as coming from the tops of certain formations. This is a simplification that usually works well enough for finding oil, but sometimes breaks down, and then is puzzling.

Reflection takes place at velocity interfaces. The contact between two formations seems like it ought to be such an interface. Usually it is, as when one is made of shale and the other carbonate. But if the two have about the same velocity, then there won't be much reflection at the interface.

Thickness is another factor. If there is a velocity difference at the top of a layer and another at the bottom of it, then each surface reflects sound. If they are fairly close together, then the reflections from the two may come to the surface one right after the other, as two separate wavelets. If the layer thins, then the wavelets get closer together. At some point, the trailing part of the upper one and the first part of the lower one coincide, blending the two into a composite reflection. As they get still closer, other parts of them coincide. Where a peak of one and a trough of the other coincide, they tend to cancel, so the amplitude of the combination is less. When two peaks or two troughs coincide, a strong peak or trough is produced. So reflection character, the combination of these individual reflection elements, changes with the thickness of the bed (Fig. 8-5). The character change can appear as a falsely dipping horizon. After the two reflections have blended, as the bed gets still thinner, only character change indicates the pinching out of the bed. Some distance before the actual zero edge is reached, the reflections are no longer separately recognizable, so the pinchout seems to occur before it actually does. Synthetic seismograms can help resolve this problem.

In seismic mapping, it is desirable to map a single layer, like the top of a producing zone. But that interface isn't alone. There are many little interfaces above and below it. And each of these reflects some energy. So the reflection is a composite of them all. If the mapping horizon is a good strong interface with a large velocity difference, it will dominate the others. But even then, the shape of the reflection is affected by elements from the different interfaces. So its character remains the same only if the various interfaces maintain their distances apart and their velocity contrasts. Changes in these quantities then, help to explain why reflection character is often fairly inconstant, and in some areas totally unreliable. It also helps explain why reflection character sometimes gives clues to lithologic changes.

With this composite nature of reflections, and the normal shape of minimum-phase wavelets, with their greatest amplitudes some tens of

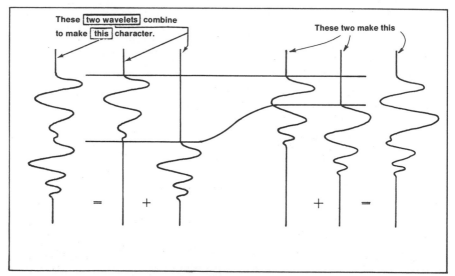

Fig. 8-5 Character changes with bed thickness

milliseconds after their origins, it is unlikely that the onset, the very start, of a reflection can be picked on a minimum-phase section. The delay varies, but is about 30 ms. So a reflection that is picked to represent the top of a formation may be a direct result of that top, but occur some 30 ms, about 150 feet, deeper. This discrepancy is usually taken care of, whether consciously or not, in selecting a velocity to make the pick fit the depth of the formation. Or if a velocity from some other source is used, then adding or subtracting a constant amount to the reflection time will make the data fit. This is not improper. It is a simple adjustment for the imperfection of the data. As long as the adjustment is constant or fits some formula, it is a perfectly reasonable empirical technique. When the processed section is zero-phase, there is no such delay, so the greatest amplitude in the reflection is at the reflecting interface.

Sometimes it may be advisable to interpret one formation by mapping another. This happens inadvertently sometimes anyway, when a not-quite-correct identification and a slightly-off velocity make a reflection seem to tie wells. An advantage of seismic data is that sections show pretty well whether the layers are generally parallel. If they are, then one can be used about as well as another for structural information about a formation. So, if a nearby reflection can be picked with more confidence than the one on the exact horizon, better information about the configuration of the rock at the desired level may be obtained from the better reflection. Notice that this applies to structural information. Lithologic information about a formation would have to be obtained from the reflection from that formation.

So, for a number of reasons, the reflection picked may not be from the formation on which information is desired. Sometimes this is misleading and puzzling. At other times it is valid and useful.

Mapping

The results of seismic interpretation are usually displayed in map form. Mapping is part of the interpretation of the data.

The maps used are the product of the surveying. Shot points are plotted, by draftsman or computer, on a base map. A base map is made on transparent material, usually film. Being transparent makes it reproducible by the ammonia-developed diazo process as transparent sepias or paper black line prints. In seismic exploration, black line prints are the standard form for both maps and sections. In addition to shot points, a base map has wells, coastlines, political boundaries, the company's leases, and other pertinent information. Then a copy of the base map is made for the reflection times of a horizon to be added to it.

Reflection times are obtained from the sections. That is, a horizon is marked on the sections, and at each shot point, or every fifth or so shot point, the reflection time for that horizon is read from the timing lines on the section. This reflection time is plotted at that shot point's location on the map. Or the time may be first converted to depth, and the depth plotted.

The plotted times or depths are contoured. The contours are lines of equal time or depth wandering around the map as dictated by the data. Then, in the simplest type of prospect—the anticlinal trap—oil may possibly have collected at locally high points in the subsurface, as shown by the mapping. That is, where the high contours are closed, going completely around some area on the map, the high area they enclose may be a good location for a well. Other situations—involving faulting, pinchouts, reefs—are more complex, but are also displayed and evaluated on seismic maps.

The seismic map is usually the final product of seismic exploration, the one on which the entire operation depends for its usefulness. In meetings with management, selected sections may be displayed with a map, but the map itself is essential to decisions about where, and whether, to drill for oil.

The process of contouring a seismic map is a matter of many choices, the contours being, except right where they cross lines of data, guesses at what would be found if there were shot points everywhere on the map. Thus the shapes of contours, particularly those farther from shot points, should not be taken too literally. The main function of contouring is just to make the numerical data more visual, so the high and low areas may be recognized quickly, without peering at individual numbers. Looking at a seismic map involves, not just seeing what structure the contours show, but also an awareness that contours near data mean more than farther ones. The farther contours had to be drawn some way to avoid leaving holes in the mapping. A refusal to contour a reasonable distance away from data would destroy the pictorial aspect of the contouring. Also, the more guessed-in contours point out leads that may need to be followed up by more lines of shooting.

Contouring—Geological and Geophysical

"Contouring is contouring, and anyone who can contour one type of data can contour another." That seems apparent, but isn't quite true. In geological contouring, the data points are precise, but usually so far apart that the regional picture and the expected shapes of features in the area must be used to fill out the data if it is to make sense.

Seismic data is not so precise, but usually consists of enough data points to establish the shapes of features. So the interpreter often doesn't need to use expected forms of features to help in guessing about how the data should be contoured.

Seismic times or depths are not as exact as well data. In a well, a formation may be found to occur at −4623 feet. But if a seismic time has been converted to a depth of −4623 feet, that is an approximation. In absolute depth, it can be 100 feet off, or more. Even in depth relative to nearby shot points, the −4623 feet may mean anything from, say, −4610 to −4635 feet. One millisecond of error can mean from 5 to 10 feet, and it is impossible to pick precisely to the millisecond. This amount of uncertainty has been called seismic wobble. So, although well data can be, and is, contoured as precise numbers, with a contour placed exactly at its proportionate spacing between wells, the wobble of seismic data would make such precision not only meaningless, but misleading (Fig. 8-6).

Some kind of smoothing is required to achieve reasonable seismic maps. There are various ways of bringing about this smoothing. Some people contour without paying close attention to exact values. That is, smoothness of contours is allowed to take precedence over the last

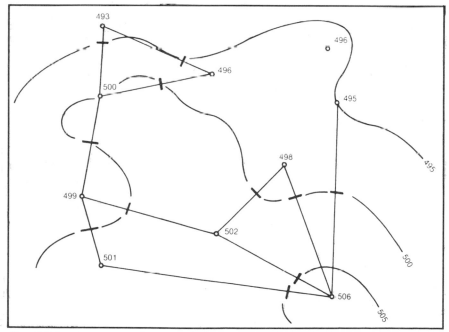

Fig. 8-6 Precise contouring

digit or so of the data (Fig. 8-7). A map so made appears to be incorrectly contoured. It isn't, but it is difficult to check.

Another way of contouring is to run a contour through a point with exactly the same value as the contour, and for numbers that aren't the same as contours, to always put a contour between higher and lower data, but to pay no, or almost no, attention to its distance from any point (Fig. 8-8). This method is somewhat arbitrary, with the numbers which happen to be the same as the contour values honored more than other numbers. The 500-foot contour has to go through a 500-foot datum. It has the advantage of being easy to check for errors, though. Just look along between two contours and make sure the data are all in the proper range.

My own way of contouring is a modification of this latter method. I assume that the value of a contour is slightly less than its label. For instance, I consider that a 500-foot contour is really about a 499.5-foot contour. Then I don't favor any data, as the label on the contour is not the same as any data point. So the 500 contour doesn't have to go through the 500-foot datum (Fig. 8-9). This is even easier to check, as all data, say, from 510 through 519, are between the 510 and the 520 contours.

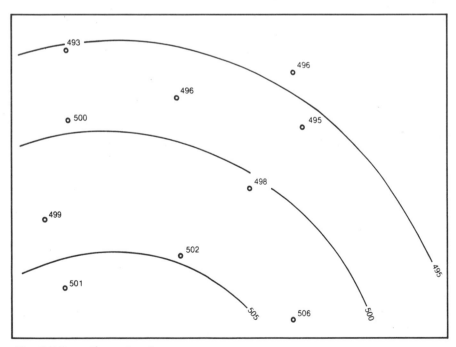

Fig. 8-7 Smooth contouring

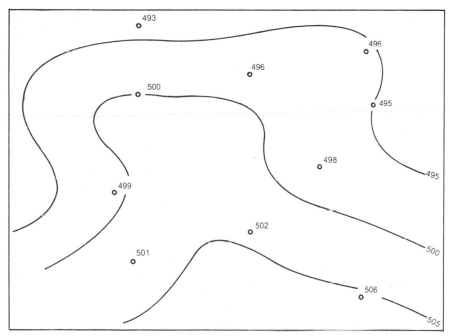

Fig. 8-8 Contours between high and low data

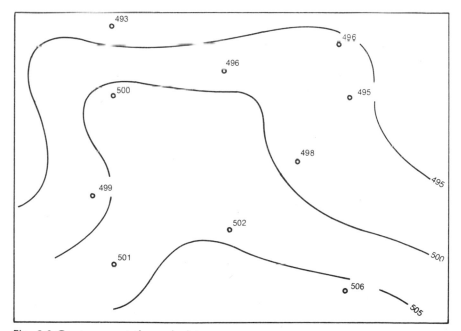

Fig. 8-9 Contours not through data

Another special characteristic of seismic data comes from the way the points are arranged. Wells are usually scattered around the countryside, except when bunched up on a producing field. Seismic data points from 2-D lines, though, are close together in long lines, but with relatively wide spaces between the lines.

In normal contouring, the pattern of blank areas surrounded by close control causes a peculiar phenomenon. The contours tend to follow the lines of control (Fig. 8-10). How did the subsurface know where the shooting was going to be?

In one area known for having many small features, we were assigned a large project. It was to be a reconnaissance program, with points all over. We shot the lines on the roads first, and were to go inside later. We shot on roads around two adjacent large square areas. What happened? The contours followed the seismic lines, making one large high and one large low (Fig. 8-11). If my contouring decisions had been different, the high and low would have traded places. You can guess what happened. When we shot inside, there turned out to be a lot of small features, no large high, no large low (Fig. 8-12).

So what is to be done with a line or two of control in an area of small features? Cautious contouring only near the lines tends to make the contours swing back and forth across them. And seismic wobble adds

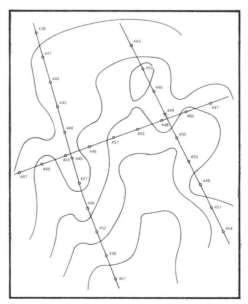

Fig. 8-10 Contours follow control

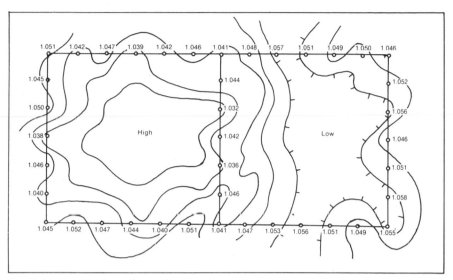

Fig. 8-11 One high and one low

to the effect (Fig. 8-13). Bold contouring, projecting trends from one line to another, becomes silly (Fig. 8-14). Drawing just short bits of contours across the lines looks like railroad ties, and is hard to understand (Fig. 8-15). Somewhat clearer is to draw little rounded beads on the lines, and hope nobody thinks they are real structures (Fig. 8-16).

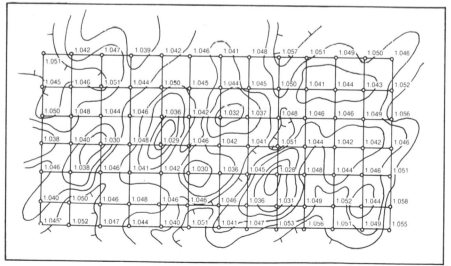

Fig. 8-12 No big features

Probably best is a compromise between the last two, that is, the railroad ties, but curved somewhat, so the contours on highs curve like parentheses (Fig. 8-17).

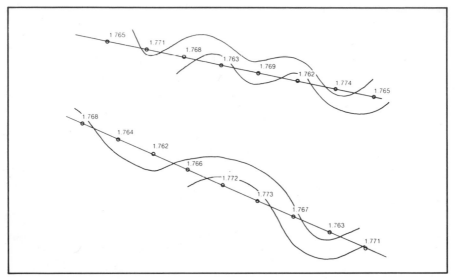

Fig. 8-13 Along control

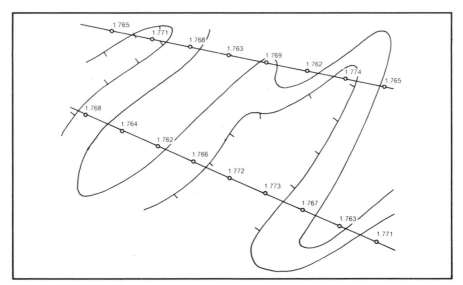

Fig. 8-14 Bold contours

Time Interval Maps

Time interval mapping is useful in several situations. It has advantages similar to those of flattening sections, naturally, as it does about the same thing, but in three dimensions rather than just two. To make a time interval map, times for two horizons are picked at each shot point; at each point the smaller time is subtracted from the larger; and these time intervals are plotted on a map and contoured.

This type of seismic map is somewhat like a geologist's isopach map, a map of the thickness of a layer or group of layers. If bedding is fairly flat, an isopach map can be made by just subtracting depths of formation tops obtained from wells or outcrops. But if there is much dip, vertical measurements do not give the true thickness. A thin bed that was folded so it was nearly vertical would appear on such a map to be thick. So a true isopach is made by correcting the vertical measurements to thickness perpendicular to bedding.

If a seismic time interval map is made from unmigrated data, the measurement is necessarily made essentially perpendicular to the lower bed—because the sound reflected perpendicularly from it. So the simple subtraction achieves, in seismic time, approximately what the dip-corrected isopach map does in thickness. Use of velocity to convert the times to distances would then make the time interval map into an isopach map.

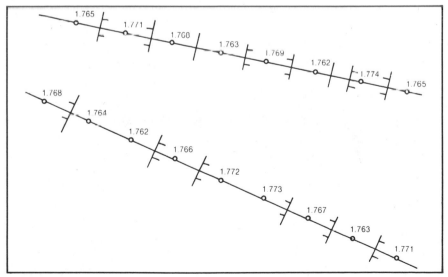

Fig. 8-15 Like railroad ties

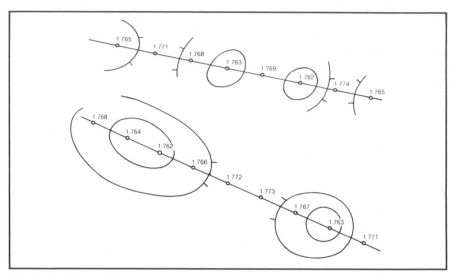

Fig. 8-16 Like beads

If migrated data is used to make the interval map, and if there is much dip, it too would require dip-correction to show true thickness.

Time interval maps are particularly useful in detecting small or subtle features in areas that have glacial drift, lignite beds, or other difficult velocity situations near the surface. If the shallower horizon chosen is one that has little dip, then not much error is made if it is assumed to be flat. In this way, the time interval map can be considered to represent the structure of the deeper horizon.

There is a terminology confusion about this type of map between Americans and Europeans. Among Americans, the reasoning is that the map is like an isopach map, the "pach" part of the word meaning thickness. So, as this map is about the same, but in time, "chron" was substituted to make isochron. In Europe, though, more correctly, "iso" for "same" and "chron" for "time" are put together to mean any map that is contoured in seismic time. A contour is a line drawn along points having the same time. I confess to having been one of those who promoted the name "isochron" specifically for an interval map. So, partly to make amends, but mostly to avoid confusion, I now speak very strictly of time interval maps and don't use the word "isochron" in either sense.

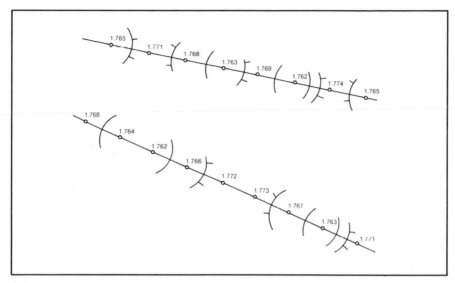

Fig. 8-17 Compromise

Other kinds of maps may be made as the occasion demands—maps of reflection character, amplitude anomalies, whatever happens to be significant in a particular area.

Misleading Features

Progress in shooting and processing through the years has improved the quality of the sections obtained, so the sections look more and more like cutaway views of the earth, with reflections having the configuration of rock layers. So, to an increasing degree, they can be interpreted by a person who is not trained in geophysics. This is one of the goals of the developments—to produce sections that can be readily understood, without a need for a specialist to explain what they mean.

But that goal hasn't been reached yet. There are a number of features that appear on a section that must be interpreted with a specialized understanding of the way they are produced. These features can mislead, and have misled, people into drilling dry holes unnecessarily.

These features have been covered as the topics were reached in the book, and they are summarized here.

Multiple reflections look like the primary reflections, but deeper in the section. They dip more steeply than the primaries, so they look like quite reasonable geologic features. They can be so strong that they override the deeper primary reflections, obscuring the deep structure. They can also make it seem that there is more sedimentary section than there actually is, by occurring on the part of the section that, for primary reflections, is below the basement. Stacking makes multiples less evident. So does deconvolution. But often multiples show through anyway, especially in the deep part of the section.

Diffractions can look like symmetrical anticlines. Or one side of a diffraction can appear to be a down-curving extension of a broken-off reflection. So, if a reflection dips down from a fault, a diffraction at the fault position can appear to be the reflection dipping in the other direction. Wells have been drilled on these apparent anticlines, sometimes fortunately finding oil trapped by the fault. Similarly, an unmigrated section may exhibit a buried focus, the "bow tie" effect of a sharp syncline. A part of that effect looks like an anticline.

A real anticline looks, on an unmigrated section, like it is larger in area than it is. So there is a chance that a well that was intended to be drilled on the flank will be off the structure altogether.

Good migration eliminates the problems of diffraction, bow ties, and too-large appearing structures (Fig. 8-18).

A dip that comes from a horizon out of the plane of the section, off to one side of the section, can, even after the section is migrated, be confused with reflections from below the line of shooting. Three dimensional migration is the most complete solution of this problem. If it is not available, or not possible with the data at hand, judgment must be applied to determine whether there are features to the side that could be recorded on the section. If dips conflict with each other, dipping in opposite directions and crossing, it may be that some are from one side.

A large feature on the surface—river bank, cliff, or the like—may reflect some seismic energy for a long distance. The appearance of this phenomenon on the section is that of a long straight dipping reflection that, extended to the surface, meets the surface feature. It may cut across other reflections. Once recognized, it can be ignored.

Velocity anomalies appear to be structural, but are caused by the sound going through masses that transmit sound at unusual velocities for the area. The masses can be salt, reefs, igneous features, gas, distorted bedding. Identifying the source of the anomaly can allow the distortion to be recognized and allowed for. In some cases, velocity data is complete enough to permit a corrected section to be made, so the velocity anomaly is removed.

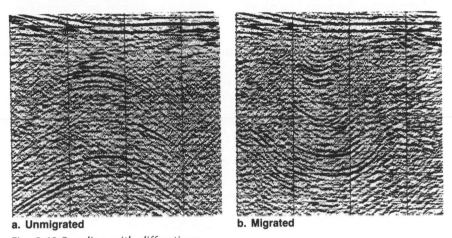

a. Unmigrated   **b. Migrated**

Fig. 8-18 Syncline with diffractions
Credit: Petty-Ray Geophysical Division, Geosource Inc.

Pinchouts are not where they appear to be on the section, but somewhere beyond the area where the two layers can be recognized.

Most seismic sections have considerable vertical exaggeration, as well as a vertical scale that does not become deeper at a uniform rate. The scale is uniform only in terms of time. The scale should be taken into account in thinking about the shape of a feature or the angle of a dip or a fault. The vertical exaggeration is generally necessary to make subtle features noticeable.

Both the vertical exaggeration and the non-uniform scale can be eliminated if velocity information is adequate, by making a depth section at a one-to-one scale. But a caution is necessary. Velocity information is usually not adequate to make a trustworthy section in depth.

Ties in Migration

A set of unmigrated 2-D sections of an area is a representation of the subsurface, but with the distortion caused by all data being shown directly below the surface points. The distortion behaves in the same way for all lines, as they are compatible with each other. In particular, they tie at intersections. Both lines at an intersection are distorted by a reflection's being moved from its actual position to a point below the intersection of the lines on the ground. So it is the same reflection on both lines, reflected from the same point in the subsurface.

In migrating, if the reflection on both sections is moved to its true position in three dimensions, then the migrated sections will also tie. But 2-D sections are migrated in two dimensions only. This migration does not treat intersecting lines so they tie. If one line extends down dip, the migration of reflections on it will move them, maybe long distances, to their correct positions. The intersecting line, if the lines intersect at a right angle, will run along strike, perpendicular to dip. It will not show the dip, so the reflections on it will not be moved at all. At the intersection, the reflection on one line will be moved to its true position, while the same reflection, on the other line, will not be moved.

The lines will not tie.

So migrating 2-D sections to get the "best" seismic data will not even let you go from one section to another. What is the best thing to do?

The simplest solution is to see if the migrated sections tie fairly well as they are. If there is not much dip, or if the lines are not along strike or dip, they may tie reasonably well. In that case, the migrated sections can be used just like unmigrated sections.

But if they don't tie, the unmigrated sections should be used to provide line ties. So the reflections can be picked on the unmigrated sections, with guidance from the migrated sections. Details of faulting, shapes of reefs, etc., will be determined from the migrated sections, and can be used in working the unmigrated ones.

The migrated sections will play a big part in drilling decisions, and may be marked and displayed to show the kind of prospect and what it looks like.

Interpretation Sequence

The actual interpretation process starts when the sections are received by the geophysicist. But results of the work are better if the interpreter is in at the start of the project, and works on the program planning, field work, and data processing. Decisions all along can have major effects on whether oil is found as a result of the interpreting.

But the interpretation starts when there are sections to work with. Areas and problems differ, but a more or less typical process of interpretation goes like this.

The interpreter looks over the sections to find the better and more continuous reflections, and then tries to identify them. If a line goes by a well that has a synthetic seismogram or VSP, or if the area is familiar, identification is probably easy. If not, then various tricks are necessary to find out what formation is represented by a particular reflection. There may be a good reflection that corresponds to the formation the interpreter would like to map. If so, fine. But if there isn't a good reflection there, then for structural control it may be better to map a nearby good reflection that looks on the section like it conforms with the formation of interest. For investigations of properties other than structure, the reflection from the formation itself is necessary.

Having selected the reflections, the interpreter starts picking. On a paper section a reflection is marked with a colored pencil. The picks are timed by reading the reflection times and writing them at the shot points. The times may be read direct from the timing lines, or from a scale. If a scale is used, it is placed in register with a nearby timing line, so paper stretch will not cause the timing to be incorrect. The reflection times may be written on the section, or tabulated by shot point number, or put on a map.

At some point, the section may intersect with another line. The picking is continued onto the other line. Going from line to line, the interpreter closes a loop by coming around to the starting point. If the two times at that point are the same, the tie is correct. If not, the mis-tie must somehow be resolved. Then there will be other problems—a loop that ties but must be undone to make another loop tie, faults with indeterminable throw, etc. Some interpreters just color the reflection, not timing any picks until all the loops in the area are tied. Others mark times on a map as they go, so they can see what is happening overall on the map, or so there can be interim mapping to show prospects or meet deadlines.

The map, after all the picking has been resolved and the times put on it, has times at shot points. Usually not all the shot points are timed, but every fifth, tenth, or so. Fault indications are also put on the map from the sections. They are marks to show where faults cross lines, and directions of throw. Amount of throw, if it seems reliable, may be added. If it is possible to correlate a fault from one line to another, that correlation may be indicated.

The map is contoured and the faults drawn in. Both contours and fault traces drawn from line to line are ways of displaying best guesses where there isn't data. So they can be trusted more where there are lines, less at places farther from the lines.

The contoured map shows places that may be prospects for drilling. A closed high feature or contours closed against a fault may show a place to investigate further with more shooting, or a place to drill a well. Non-structural features may be added to the map or put on a separate map. Bright spots, dim spots, flat spots, reefs may be prospective.

Interval maps may be made by subtracting the times on one map from those on another. These maps can show reefs or pinchouts, or substantiate the validity of the horizon maps. If near-surface conditions are difficult, the interval maps, being free of near-surface effects, will not have the problems caused by them.

Anomalous features, especially those recommended for drilling, are discussed with geologists to consider them in the light of both disciplines.

The results of the interpretation are shown to management in meetings on further action to be taken in the area. It may be decided to abandon the area as not looking prospective, to do more shooting on it, to drill on it, or to wait a while before doing anything.

Whatever is done, the things learned in the area are preserved in a report. The report usually contains a number of pages bound together. It contains such information as: dates and other details of shooting, steps in processing, way the interpretation was carried out, results of the interpretation, and recommendations for further action. This report is kept in the company files, to be referred to if a re-interpretation is to be done, or more shooting, etc.

Interactive Interpretation

Seismic interpretation, for most of the years that seismic exploration has been in existence, has been done pretty much the same old way, with people making pencil marks on records or sections, reading the reflection times of those marks, plotting the times on maps, and contouring the maps—all by hand. But 3-D data is so voluminous that interpreting it in the conventional way would require great stacks of paper. There are many parallel lines in a 3-D survey. And by selecting one trace from each of these lines, a section at right angles to them can be displayed. A whole set of those sections can be made, or sections in any other direction. And the data can also be displayed as horizontal slices. To effectively use all that capability would increase the printing bill and the storage and filing problems a great deal. And the interpreter would be engulfed in paper.

So 3-D data needs to be handled in a different way. Most of those sections don't need to be printed, if they can just be looked at. Some mechanical devices for 3-D interpretation have been developed. Many sections or slices can be put on frames of movie film and displayed as a movie or projected one at a time to be worked on. And there are boxes for viewing several transparent sections at a time. With one of these devices or just a series of prints, contouring can be done on time slices by tracing a reflection on one slice, then the same reflection on another slice, etc.

But the devices are mechanical, and can't handle masses of data as well as a computer can. A number of interpretation systems that use interactive displays on computer screens are in use. An interactive computer system is one that interacts with a person in real time (Fig. 8-19). That is, the person gives commands, and the computer responds quickly. These systems are still developing, but even at their present stage, they make interpretation many times as fast as it was, and much more correct at the same time. They are well adapted to the interpretation of 3-D data.

For an interactive interpretation to begin, the data for the area is put in the memory of a computer. New data is usually obtained in a form that a computer can use. But, for it to be effectively fitted in with continuing projects and other things the company has already done, it must be compatible with them.

Putting a mass of information in computer form so the different

Fig. 8-19 Interactive interpretation
Credit: Western Geophysical Co. of America

parts are compatible is called building a database. A database, once made, will be used for a long time. Any inconsistencies in it will cause trouble 'time and again as the data is worked on. So at some time a company may need to engage in the large and careful project of building a database. Maps of different seismic programs, geologists' maps, and maps of leases and land ownership must be made to fit together. A great backlog of older map data in companies' files must be converted from hand-drafted maps to digital form. This digitizing can be done by touching the various positions on the map with a pen-like instrument attached to a computer. So, either as a large project or doing each area as needed, there is a lot of digitizing to be done, to use an interactive system effectively.

Interactive systems have a number of different functions that they can perform, but not necessarily all the functions are included in any one system. And each company that is developing one is trying to make it as complete as possible. So rather than describing a specific system, I'll just give the main features that may be found in various ones of them.

The interactive station is a terminal of a shared computer, or it is a stand-alone unit with its own computer. There are at least two color CRTs—television-type screens—and perhaps a third. There is a type-writer-like keyboard (Fig. 8-20).

With this setup, and with the data for the area in the computer's memory, the interpreter punches commands on the keyboard to start the program. A menu appears on a screen, offering choices to which the interpreter responds. For instance, the menu may display several things that can be done, with instructions to punch certain numbers or letters for each. The interpreter may thus indicate that a section is to be displayed on the screen. Then the menu may change to a request for the line number. The interpreter punches that, and is asked for the start and end shot points to be displayed. This goes on through type of display—whether VD or wiggle-VA—attributes to be shown in color, choice of color code, etc. There may also be some keyboard commands that do not appear on the menu, but have to be memorized. There may be a cursor, a bright spot on the screen that moves in response to commands.

A mouse is a hand-size object the interpreter can move around on the desk top. A cursor on the screen mimics the mouse's motions. This allows the interpreter to "point" to choices on the screen by moving the cursor to them, and also to point to places on a section or map that is displayed.

Fig. 8-20 Using interactive system
Credit: Petty-Ray Geophysical Division, Geosource Inc.

The interpreter can use a graphics tablet, a smooth board covered with symbols, and a pen connected by wire to the computer. The symbols take the place of some or all of the menu. A selection is made by touching the pen to a symbol. In whatever way the commands are given, there is great freedom from the physical handling of data.

Although the systems were designed largely for the special requirements of 3-D interpretation, they apply well to 2-D data also. To work on a vertical section, from either a 2-D or a 3-D area, a part of a line is called up. A reflection is selected by moving the cursor to it. Then, it can be picked by moving the cursor to various points along it, and instructing the computer to draw straight lines between the points. Or if the reflection is a good one, the computer can be told to pick it. It will draw a line along the reflection. Where it goes wrong, the interpreter can hand-pick through the confusing place, and then have the computer pick beyond there.

Vertical sections can be displayed at different scales, to show the overall picture or fine detail enlarged. The vertical to horizontal scale can be varied, horizontally compressing the section to the desired degree. Horizons can be flattened to show the sections as of some earlier geologic time. A section can be "cut" apart along a fault plane, and the parts shifted to remove the throw of the fault (Fig. 8-21). This permits the interpreter to correlate more reliably across the fault. A growth fault must be shifted repeatedly to make the different reflections match, so throws at those levels can be determined.

In 2-D work, each line is shot and processed more or less as a separate entity, so there are problems with ties at line intersections. Parts of two lines can be shown on the screen, joined at the intersection. Interpreter or computer can pick across the intersection. Going on around a loop can be made easy by displaying the parts of sections that

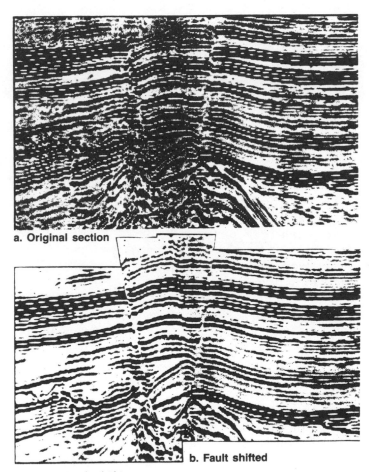

Fig. 8-21 Fault shifting
    Credit: Western Geophysical Co. of America

make up the loop as one straight section. A display of mis-ties at inter-sections can be made on the other screen—in map form, or as a list.

With a section on one screen and a base map on another, the computer can read the times of picks and display them on the map. It can also contour the map (Fig. 8-22). Or the picks can be displayed as a perspective representation of the highs and lows, looking like a net draped over an invisible structure (Fig. 8-23). This representation can be rotated to be viewed from different sides. Even a personal computer can run some contouring programs, producing both these types of display.

Other functions of the interactive system are most applicable for interpreting 3-D data. After a line is picked, and the next parallel line displayed, the picks from the first line can be displayed on the new one. They will probably be almost on the same reflection on the second line. Then a little adjustment can make the picks fit the second line. This allows the interpreter to go through a set of sections quickly with

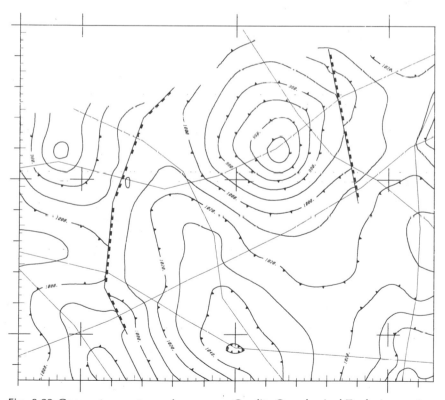

Fig. 8-22 Computer contoured map   *Credit: Geophysical Techniques, Inc.*

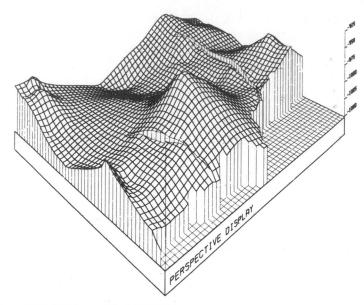

VIEW = 225, TILT = 35 DEGREES
Fig. 8-23 Perspective view   *Credit: Geophysical Techniques, Inc.*

compatible picks. A contoured map on the other screen can be growing as each new line is picked. In 3-D work, since the data is acquired in
a coordinated way and processed as a unit, there is not likely to be a
problem in relating one line with another.

A time slice can be displayed as easily as a section. Picks on the slice
can be made manually or by computer just as they are on a section. For
interpreting from section to slice, a combined display can be shown.
The two may be joined as though prints had been laid on a flat surface
and taped together. Or they may be displayed as a perspective view,
with the slice appearing to be flat and the section hanging down at the
front edge of it, or standing up at its back edge. With this kind of
display it is easy to interpret the relationships between the two. A fault
trace on a slice and the throw of the fault on a section can be interpreted together.

There are more combinations of displays. A corner of a cube can be
shown, with its top a slice and the sides being sections. Narrow horizontal strips of several sections can be displayed one above the other in
an "accordion" display, for comparison of a reflection. And so on, limited only by people's ingenuity.

A sequence of slices or sections can be displayed (Fig. 8-24). This
viewing of the whole block may be done at the start of an interpreta-

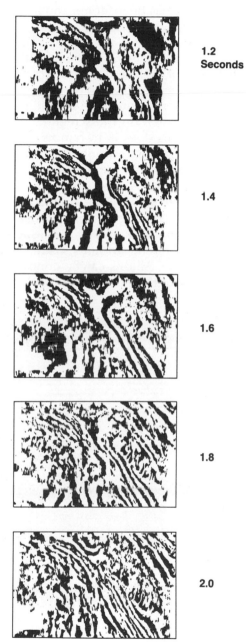

1.2
Seconds

1.4

1.6

1.8

2.0

Fig. 8-24 Sequence of time slices   *Credit: Geophysical Service Inc.*

tion, to get an overall view of the geology. It can also be useful later to check on interpretation that has been done, and maybe change some of it.

After an interpretation is made on an interactive system, there may be a need to put some of the results on a wall, show them at a meeting, or mail them somewhere. For these purposes hardcopy, that is, prints, can be made directly from the information on the system. A color plotter connected to the computer can produce a section, slice, cube, map, etc. so it looks like the one seen on the screen (Fig. 8-25).

The main thing that all this does is to remove from interpretation most of the time-consuming filing, sorting, searching, reading times, and plotting. It allows interpreters to spend more of their time interpreting. All through the process of interpreting by hand is the frustration of either not being able to make all the comparisons that would be useful, or of the interpretation taking too long. With a good interactive system, an interpreter can run back and forth through a collection of sections, compress them different amounts, flatten horizons, look at different attributes of the data. All this can be done so rapidly that the interpreter won't lose track of the question in mind while bringing up

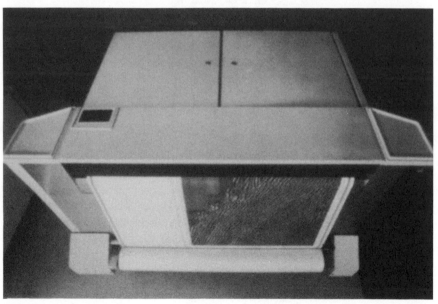

Fig. 8-25 Color plotter
Credit: Petty-Ray Geophysical Division, Geosource Inc.

data. That makes it much more likely that the interpretation can be thorough and well thought out.

These wonders, of course, depend on having the data processed and put in the computer. And it depends on the availability of an expensive setup for the interpreter's use. If there are several interpreters to use one interactive device, then each one may have to wait for a turn with it. The ideal arrangement is for there to be a terminal or a system for each interpreter, so it is idle until the interpreter needs it, and then immediately available, perhaps with the data already loaded in it. The systems are also subject to power outages and equipment failures. But the speed and quality of work done with them makes it well worth while to allow the interpretation to be interrupted while such problems are resolved, and then continue interpreting.

Forty years ago in seismic exploration, "computer" meant a person who calculated corrections for seismic records. About that time, the first electronic computers were built. They filled rooms. The calculations they could perform can now be done by small, cheap pocket calculators. Mentally projecting that rate of change into even the near future is dizzying. It is possible that the first rush of development is over, and the rate of change will slow somewhat. Even if that is true, what is likely to happen to seismic interpretation?

As the minicomputers become more powerful, the trend to using them instead of connections to mainframe computers is likely to continue. More power in the minicomputers will mean that their memories will hold more data, and that they will be smaller and cheaper. For one thing, the big television-type screens will surely soon be replaced by one of the several types of flat screen now in limited use. That will make it easier to work on a section, slice, or map than the present vertical screen in front of the interpreter. A flat screen might be held in the lap or laid on a table where touch-screen techniques would be convenient.

Systems will also surely acquire new abilities, making or suggesting more of the interpreting decisions, maybe using a partial artificial intelligence in the form of stored geological information on how rocks and reflections behave. Farther in the future, systems may be briefcase-size and battery powered, perhaps projecting the sections and maps onto convenient walls or desktops.

CHAPTER 9

Interpretation Three Ways

In the preceding chapter the methods of producing a map from processed data were discussed. Now this will be expanded into three different approaches to interpreting seismic data.

Seismic exploration is often said to be of two types, structural and stratigraphic. "Structure" and "stratigraphy" are good geological terms that clearly express geological concepts. But the things seismic exploration determines about the subsurface do not naturally separate into structure and stratigraphy. I'll describe three divisions of my own that I think better fit the way seismic exploration separates naturally into different types.

The ups and downs of a reflection show the shape of some horizon in the subsurface. It may be the top of a formation. It may have been bent by tectonic forces, and, where those forces were too great for its plasticity, broken. In those places the reflection is following structure. But a reflection may also go over the top of a reef or sand buildup. That isn't structure, but it looks just the same on a seismic section. So these characteristics are better described as configuration. Traditional seismic mapping depicts configuration in the subsurface—the tops of masses of rock, whether they were shaped by tectonics or originally formed that way.

Another approach to hunting for oil is to investigate how rocks were deposited, and so is called seismic stratigraphy. The feature of seismic data that is investigated is, not the shape of a single horizon, but the relationships of sets of reflections, one above the other. This is a matter of how a group of reflections conform, call it conformation.

A third approach uses qualities of the reflected energy to determine the makeup, the constituents of rock, rather than shape. This approach is not concerned with the path followed by a single reflection or a group of reflections, but with the nature of the reflections and how they respond to differences in lithology. This includes investigations of reflection character and of seismic attributes—velocity, frequency, amplitude, polarity. And in particular, it involves using different offsets of receiver from source to detect rock properties. This last is called seismic lithology.

So the natural division of seismic exploration seems to me to be the determination of configuration, conformation, and constituents of rocks, that is, how they're shaped, how they were laid down, and what they're made of. To describe interpreting activities as they are normally performed, these three divisions will be used in this chapter.

But people conduct seismic investigations to find oil, not to work in some category. So all three of the uses may be applied to any area. It is necessary to know the configuration to find high places, because no matter how oil is trapped, it still floats to the top of water in the trap. The depositional history leads to ideas of how oil might be trapped in the area, and may indicate traps directly. The mineral makeup and content of the pores in rock are needed to determine which reflections are from porous rocks, and something of the trapping mechanism to be expected.

So it is important to realize that the three cons are all interrelated, and an investigation of any one of them may use either or both of the others to solve its problems.

Configuration

"How they're shaped"

The configuration use of seismic exploration is traditionally and primarily for finding locally high parts of formations in which oil or gas may be trapped—the top of an anticline, fault block, reef, sand buildup. Some of these features are shapes that were formed and later buried, others were caused by tectonics acting on existing features. Configuration also finds features that are the result of both these processes together—tilted reefs, faulted sand bars, etc.

Large Structures

Large structures, either high-relief or broad with little dip, have their own subtleties that keep them from being as easy to find as it might seem. The broad low-relief structures can be spread over so much territory that seismic programs may be shot on them, without ever detecting the structure because they don't extend off of them. This happened to a friend of mine in the Scurry Field of West Texas. Shot an area that had been assigned on the feature, but none of the lines extended beyond it. The field was discovered later by someone else. Similarly, a prospect may be on what seems to be regional dip, but is the flank of a large structure. And the low relief on a broad structure may be difficult to notice, because it is too flat to be apparent on the sections.

The steep structures are necessarily smaller in area. If they were both big and steep, they'd run out of room underground, and have to stick up in the air. Some do, of course, for instance the "sheepherder anticlines" of Wyoming, so named, I would guess, because the sheepherders knew about them before the oil business got interested.

These big structures, particularly when they don't show on the surface, can be subtle because they are so steep. For energy to strike a steeply dipping bed perpendicularly, so it can be reflected back to the shot point, the shot point must be far from the structure. Then the dip looks out of place as it conflicts with flatter reflections from nearby, and so is easy to dismiss as noise. Right over the structure, the turnover at the crest is less steep, so it is easier to recognize, but that part may extend only a short distance (Fig. 9-1). One prolific structure in Wyoming is producing because a seismic contractor saw one strange dip, and insisted that it be investigated further with more shooting.

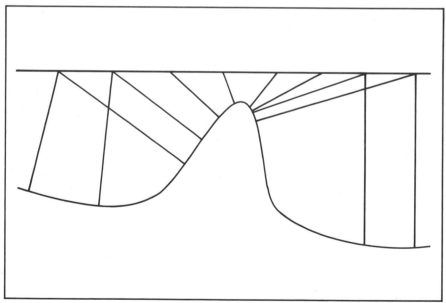

Fig. 9-1 Steep structure

Looking for these steep, narrow structures calls for long enough lines to detect the steep parts, an open-minded attitude toward stray dips, and good migration of the data.

Small Features

Small structures, and small features of any other type, have some special exploration requirements, just because they are small.

Exploration has grown up based on a search for large anomalies. And in a new area it tends to grow again in the same way. Of course bigger fields seem better to find, but structures smaller in area can often be well worth seeking too. Some of the world's largest fields in terms of production are small in area. Because of the large-anomaly bias, small fields tend to be looked for with big-field philosophy. This is like trying to catch minnows with a salmon net.

Exploration for small fields calls for some special techniques:

Seismic lines should form a fine enough mesh to find them. In some areas this calls for lines a quarter mile apart. Widely spaced lines are almost useless for this purpose, giving leads only to those features they happen to cross, and providing little information about them.

A well based on the seismic data should be drilled only on a seismic line. Off to the side of the line you can only guess that the location is on the feature.

From this it follows that the lines should run through locations that can be drilled. It is a waste of money to shoot the lines on roads in an area in which drilling close to roads is not allowed.

After the first shooting, detail lines may be required anyway, to locate a well. So, instead of looking first in places you know you won't be allowed to drill, the original search should be conducted with lines through legally drillable locations.

This is an economic restriction, so it does not apply to the same extent to more economically obtained data. Seismic data purchased through data exchanges, traded for, or included in a farmout deal can form a network on which further shooting is based. However, the additional shooting should be used not only for detailing leads found in the old data, but can also fill in the spaces between lines. This can be in stages, of course, first following up the leads. The larger gaps in control, though, should not be considered unprospective, and should be checked when economic, lease, etc., conditions permit.

Some areas of this type have had some wells drilled on them, and have some production, but the dry holes have discouraged the industry. From experience in seeing different companies operate in such an area, a pattern is evident. A company starts with the large-structure assumptions—that reconnaissance lines should be shot on roads for economy, that fairly widely spaced control is adequate, that a well can be drilled off of the control as long as it is on the contoured structure. The company either drills a few dry holes and pulls out; or is lucky at first, finds some oil, and starts to learn the lessons of the area. At any point, it can fail to learn, and pull out. So a number of dry holes drilled by various companies do not condemn such an area. They just indicate the companies that used the wrong approach. If, in an area of small structures, you hear of a dry hole drilled on a structure, the next questions to ask are whether the structure was detailed sufficiently, whether the highest point was drilled, and whether the well was drilled on a seismic line or based on contours a quarter mile or so away from the line.

In any area of small features, a 3-D survey will solve most of the problems described. But the cost of 3-D shooting does not usually permit its use for general exploration. But sometimes it is worthwhile to shoot 3-D on a prospect just before drilling it.

Salt Solution Features

In the salt-solution areas of the Williston Basin, many small fields have been found and surely many more will be found. They require a special outlook and technique.

In the geologic history, a local part of a layer of salt dissolved, allowing later deposition to fill in, becoming thick where the solution had taken place. Then, after more deposition, the rest of the salt layer dissolved, so the overlying beds sagged. The thick spot where the first deposition took place caused other beds to drape over it, creating a trap. The features resulting are small, say from 160 to 640 acres in area. But there can be a number of them, in a small area, so they are worth looking for. Exploring for these features requires close seismic control, like any other small features.

Prospects are found largely by time interval mapping. The beds under the feature are undisturbed, and over it there is drape, which diminishes toward the surface. So intervals from whatever is the pay zone up or down will show the features by thickening or thinning.

The near-surface situation may be complicated by glacial drift. This is not a great problem, though, as any feature found on a structure map can be checked on an interval map.

Salt Domes

Salt domes are the key to petroleum accumulation in many areas, including the U.S. Gulf Coast, North Germany, parts of the Canadian Arctic. Salt dome exploration was the first common exploration use of seismograph. Refraction lines were shot so the energy, where it came in more quickly on some lines than others, indicated a mass of salt. Reflection shooting is now used to find salt domes and determine their shapes.

A round topographic elevation on land or on the sea floor can be a first indication of a salt dome. On land, these features are easily found, and then can be shot in detail, with no need for reconnaissance shooting. At sea, though, they aren't so readily seen. The usual approach is to shoot a grid of, say, 2 by 3 or 3 by 4 miles. The salt domes located by the grid are then closely detailed. With a good pattern of detail lines, a good picture of the shape of the dome can be obtained.

On a seismic section, a salt dome is an area without the normal reflections, often extending up from the bottom of the section. Horizons may be bent upward as they approach the dome. Those above may show drape over it (Fig. 9-2). On a time slice, the area without reflections appears as a round blank spot (Fig. 9-3).

Making a good picture of the dome, though, does not solve the problem (Fig. 9-4). There is still drilling to be recommended, and the shape of the dome and the beds over and around it are not sufficient information to determine the best drilling location. Oil may be trapped in:

A bed upturned against the dome,
A pinchout in an upturned bed,
Drape over the dome,
A fault trap,
A bed upturned against a shale sheath on the dome,
Under an overhang of salt.

So there may be oil on any side of the dome or directly over it. On the flanks many depths may need to be tested.

Salt, shale, and igneous intrusions can look alike on sections. Any of them can form oil traps, in the same ways. If there are good reflections in or under them, they can be distinguished from one another by velocities derived from normal moveout, as described in "Velocity Analysis" in Chapter 5, Data Processing. The amplitude anomalies given under

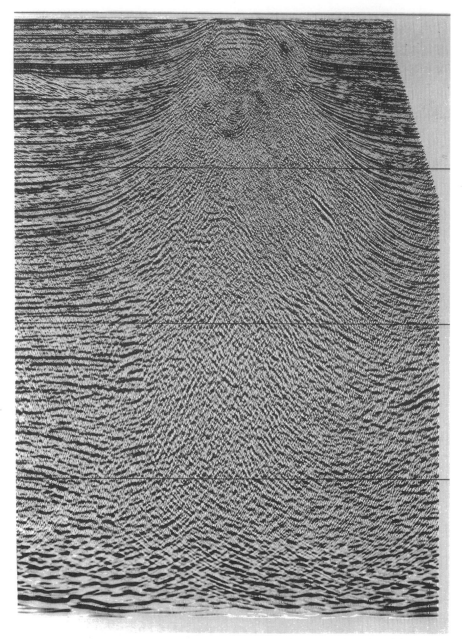

Fig. 9-2 Salt dome section   *Credit: Western Geophysical Co. of America*

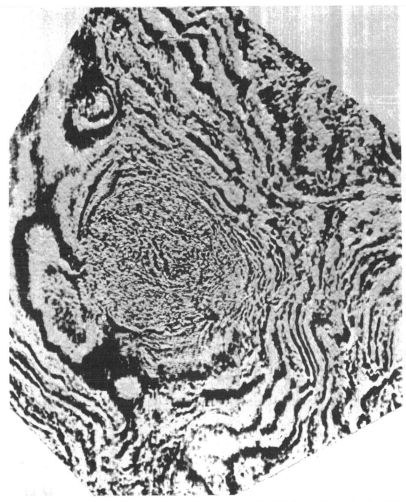

Fig. 9-3 Time slice of salt dome   *Credit: Digicon Geophysical Corp.*

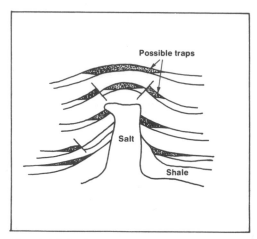

Fig. 9-4 Salt dome traps

"Constituents" later in this chapter are an aid in finding the gas and therefore oil associated with salt domes.

Reefs

Reefs can appear in very different ways on seismic sections. They can be clearly visible, subtle and difficult to delineate, or totally invisible. These various degrees of detectability can occur near each other, even in adjacent reefs, although more often they are typical of different areas. So the reefs can be detected directly, indirectly, or not at all.

Seismic reflection takes place at velocity contrasts. The limestone of a reef, if it differs in velocity from the rock over and around it, will produce a reflection. Some reefs, probably most, are overlain by sand or shale of considerably lower velocity than the limestone. Or the overlying material may itself include some carbonate, and have a velocity close to that of the reef. A reef containing gas will have a lower velocity than limestone with water in its pores, and so may have a velocity near that of the overlying rocks. The greater the contrast, the stronger the reflection, so these various situations can lead to all degrees of reflection strength.

The well-defined reefs, for instance some in Indonesia, can easily be seen on sections, and can be easily mapped as to area and upper surface. Mapping of interior features, layering, etc., is generally beyond the present ability of seismic exploration. Thickness of the reef can be determined only if the reef is thick enough, there are good enough reflections at both top and bottom, and the reef is shallow enough, to make a good velocity determination from normal moveout, so thickness can be calculated.

The less-clear reefs may in some cases be the more productive ones. A reef with open spaces filled with gas will have a slower velocity of sound than one containing water, and so may not contrast as much with rocks around it, and may not produce as strong a reflection. The subtlety can be such that a reef is clearly seen on one section, and not visible at all on a nearby line, or even on one intersecting the first on the reef.

When the reef itself does not show up on a section, there may be indirect evidence that will indicate its presence, and even its outline and thickness. The carbonate reef material is usually not as compressible as the surrounding formations, so differential compaction causes these overlying rocks to be depressed around the reef, but not over the reef. This drape is greatest in layers closest to the reef level, so mapping a reflection just above the reef level can indicate, by its locally uplifted places, the presence of reefs below it.

In a fairly thoroughly explored area, a numerical estimate of reef height can be obtained from the drape. This is a matter of experience in interpreting seismic data and data from wells drilled into a number of reefs in the area, and is only a rule of thumb. So many milliseconds of drape may represent some thickness of reef. This relationship won't normally stay constant over a large area, and so may have to be worked out separately for various parts of the area.

In an area of fairly flat horizons, the interval maps are valid in themselves. But, looking for reefs in a tectonically disturbed area, a structure map may also be necessary to determine the tilt of the reef, so a well can be drilled in the highest part. Spilling out of the oil from the upper end of a steeply tilted reef is prevented only if the material underlying the reef as well as that above is impermeable.

In most cases, the reef velocities are higher than velocities in the surrounding rock. This causes the phenomenon called velocity pullup. A reflection from a layer below the reef may show some apparently high structure directly under the reef. This occurs because the sound, traveling down to the layer and back, accomplishes the trip more quickly where its path is through the reef than at other places. The pullup is not real structure, but is useful as a clue to the reef. And, of course, an apparent structure under a reef does not have to be pullup, but could be an uplifted part of an old seabed, on which the reef grew.

When the surrounding rock has the same velocity as the reef, then there is no velocity anomaly. Also, when the velocities are the same, the compressibilities are about the same too, so there may not be any drape in the area. This makes such reefs difficult or impossible to find with present-day seismic exploration—naturally, as velocity differences are what seismograph depends on to yield any data at all.

In areas where the reef velocity is slower than the velocities of other rocks at the depth of the reef, the pullup phenomenon occurs in reverse. A horizon will seem to sag under the reef. This pulldown is just as good a clue as pullup. Pulldown also occurs if there is gas in the rock, slowing the reef velocity.

Reef indications tend to vary from place to place, so it is best to have some experience in the particular area. In an area that is fairly new to oil exploration, but with some oil already discovered, then it is very important to get at least a line of seismic data over a discovered field in the area. The data can be obtained by trade or purchase, or by shooting it. Failing this, well data can help somewhat in deciding how a reef might appear on a seismic section.

In some areas, a reflection may be interrupted by reefs poking up through it. So the local absence of a reflection may be a reef indication.

A thorough velocity analysis of a section may give additional information on velocity differences between reef and non-reef. Velocity data comes from reflections, so this depends on there being a strong reflection at the top of the reef and one at its bottom. However, reefs are usually not very thick, say less than 1000 feet thick, so there is not much effect on the overall velocities to the horizons, and the velocity data may not give very solid information.

Geological processes related to the reef, either as it grew or resulting from its presence, may influence velocities above it. Differential compaction is one such process. There can also be differences in cementation, etc. So velocity investigations of zones close above the reef may help detect it.

These same processes may make layering thinner, or anyway different, over the reef, so analyses of seismic frequencies may be useful. They can be displayed in color, superimposed on the section.

Reflection character, influenced as it is by thickness changes and lithologic changes, may be anomalous in horizons over reefs. So careful examinations of character may help locate otherwise elusive reefs.

Pinnacle reefs are small in area, and so have the special requirements of small features. They call for closely spaced lines, and for wells drilled at high points that are on seismic lines.

Faults

Faulting is one of the more important factors in the configuration of horizons. But on 2-D lines faults are very difficult to map correctly.

Under good conditions, a fault can show up clearly on a seismic section as offsets of reflecting horizons, with the breaks on the various horizons following a slanting path on the section. This path represents the fault plane as it intersects the seismic line. Faults are clearest on migrated sections (Figs. 9-5a & -5b).

But there are problems concerning the angle of the fault and its direction. And, when faults are more subtle, or the section is poorer, the faults are hard to find on the section, and throw is difficult to determine.

There are several kinds of less direct evidence for a fault, when it isn't clearly visible.

A loop may not tie, because a fault was not seen. It can be sought all around the loop.

A jump in correlation may occur, without a break in continuity. That is, a horizon may seem to be continuous, but two places on the line may appear similar only if that continuity is not followed (Fig. 9-6). This can be pretty subtle. It may be that the bed is faulted, but its reflection just happened to line up with another reflection at the break. Or there may not be a fault, and the change of character indicates a lithologic change in the formation. It is a matter of working with the sections in the area long enough to become so familiar with them that you can come to a conclusion with some confidence.

A change in the dip of one horizon or a number of horizons, may indicate faulting. As an example, suppose on part of the line everything is flat; then, at some point on each horizon, there is a change to some amount of dip, and the points of change on the different horizons are aligned on a slant. This may mean that there is a fault plane at the slanting alignment, or it may be just what it looks like, a change of dip along a hinge line. A clearer fault indication is for the dip to change, and then farther along the section, to change back to its original attitude (Fig. 9-7). Again, the section could just be telling the truth. My own preference, when record quality is good and there is no other evidence to the contrary, is to consider that these situations indicate dip only, not faults, and that the rock layers are bent but not broken.

Any non-vertical break in the appearance of the section—in continuity, character, anything—may be caused by a fault. Non-vertical,

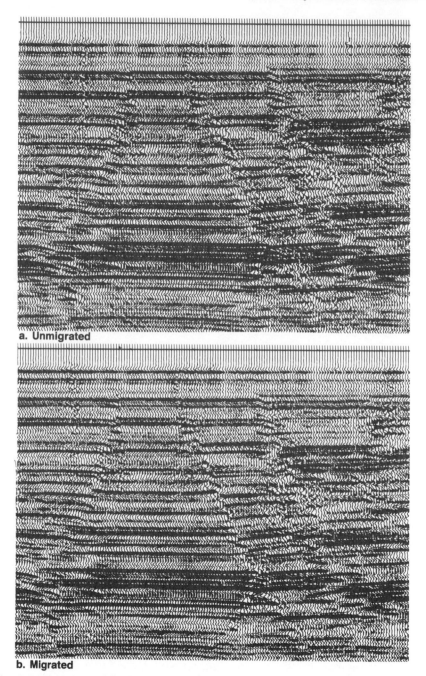

a. Unmigrated

b. Migrated

Fig. 9-5 Migration of faults—look at from one end

Credit: Western Geophysical Co. of America

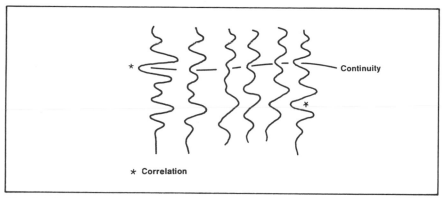

Fig. 9-6 Continuity vs. Correlation

because a vertical break may be only a result of a change in the near-surface, or a processing problem, or even a place where two segments of seismic section film, with slightly different processing (or even different photographic developing of the film) have been spliced together. There can of course be vertical faults, but these possibilities should be considered before interpreting the vertical breaks as faults.

Local distortion of bedding, in the form of drag against a fault, may be visible on the section.

Diffractions are a good indication of faulting. Where a formation is broken off by a fault, there is likely to be a diffraction on the section. Usually only half of the arc of the diffraction is apparent, the part beyond the end of the formation. So the layer seems not to end, but to curve downward, appearing to be folding of the layer. If the diffraction cuts across other reflections, then it may be apparent from the superimposed look that it is a diffraction. At a large fault a number of formations may be broken so there can, in good circumstances, be a slanting alignment of diffractions (Fig. 9-8). Migration, in eliminating the diffractions, will make the fault more clearly visible. But the diffractions are fault indications too—ones that can be seen only on the unmigrated section. So a comparison of the two types of section may be useful in finding faults.

Non-seismic evidence can also help in fault interpretation. If a well cuts a fault, and the well is on a seismic line, then it is saying that there is a fault at that point. If it can't be seen on the section, there is a problem, but usually the well logs and section can be worked on together and a joint solution arrived at. Either may cause the other to be reinterpreted somewhat. A fault in a well that is not on the line, but only nearby, is less direct evidence, and requires more flexible treatment. Surface faulting may also be tied in with seismic sections. In some areas it is advantageous to plot surface faulting and other surface geology along the top edge of the section.

A caution is in order in seismic interpretation of faults. In areas known to have faults, interpreters sometimes feel that it is necessary to map faults, even if there is little or no evidence on the section. However it is sometimes better to omit faults, and include a note on the map

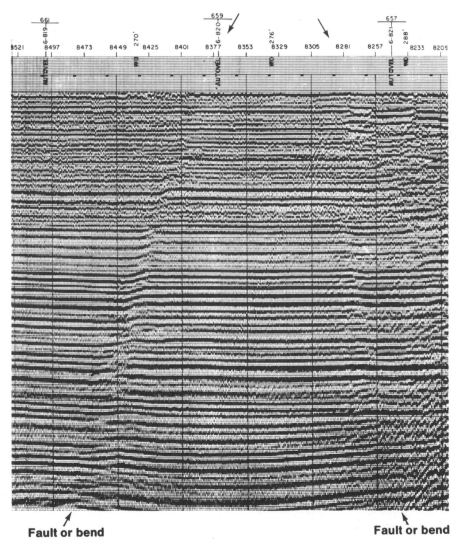

Fault or bend   **Fault or bend**

Fig. 9-7 Possible faults   *Credit: Teledyne Exploration Co.*

to the effect that there are probably many faults, but that they are not visible on the sections.

Where a seismic line crosses a fault, and the fault shows clearly on the section, as a slanting line of broken or definitely offset reflections (Fig. 9-9), and if the seismic line crosses the fault at a right angle, then the angle of the fault seen on the section will represent the true angle of faulting. It will represent it in terms of seismic time as the vertical dimension and distance as the horizontal. To determine the angle of

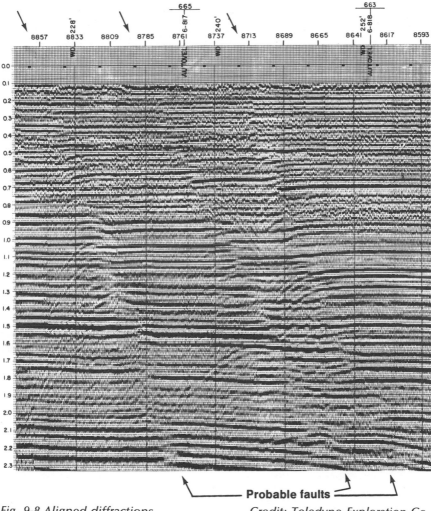

Fig. 9-8 Aligned diffractions   *Credit: Teledyne Exploration Co.*

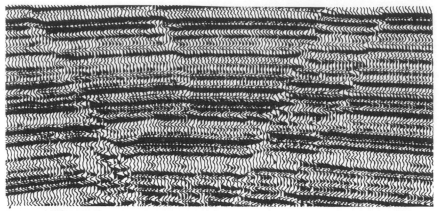

Fig. 9-9 Clear faults
   Credit: Western Geophysical Co. of America

the fault, use whatever velocity data is available to calculate the depths of two or more points on the fault. Then, using the locations of those points, obtained from the shot point numbers at the top of the section and their distance apart (measured on a map if the geophone groups weren't uniformly spaced), plot the points on another piece of paper at a one-to-one scale.

If the seismic line is the one on which the fault was first discovered though, it is unlikely to cross it perpendicularly. If it is a detail line, shot after the fault was found and mapped, then it could be planned to pretty well cross it at a right angle.

In any other case, the line crosses the fault at less than a right angle, more obliquely. The fault appears on the section to be less steep. The ultimate of this is when the line runs parallel to the fault, so the fault plane would appear horizontal if it could be seen at all. The general case is that an apparent fault plane on a section, after being plotted at a 1:1 scale, is less steep than the fault itself.

The mapping of faults is probably the most severe problem in 2-D seismic exploration. A map of a drilled-up salt dome oil field may be largely a crisscross pattern of faults. A cross section of the field may show the same type of fault pattern. The patterns are derived from studies of well logs, production histories, etc., and are themselves fallible, but represent the picture that was obtained with much data and effort by skilled specialists. Even if the faults aren't just as shown, the map and section indicate that the field truly is broken up into a hodgepodge three dimensional pattern of fault blocks.

Then, if a seismic survey is run over a similar prospect, you can be almost certain that a similar complex of faulting exists there. From that it is normal to assume that the faulting should be mapped.

Fault evidence occurs on a seismic section, which is two dimensional. If it shows clearly, the fault trace is a line of broken beds slanting across the section. In a highly faulted area, there should be a number of these fault traces on a section. Then, on the next section, say a nearby parallel line, there should also be a set of faults. Now comes the interpretation problem. Which fault on the first line connects with which one on the second line? Occasionally, a fault may be recognizable— the biggest fault in the area, or the only one downthrown to the southeast, for instance—but a common situation is that there is no way to identify a fault from line to line.

Something has to be done in order to make a map. It won't do to just indicate where faults cross a line. If it is left at that, the interpretation isn't finished, and someone other than the interpreter has to complete the interpretation by guessing how the fault indications connect. The interpreter has to make the guess, and may be able to guide it somewhat by an idea of local geology. Then, when the map is complete, there is an interpreted pattern of faults on it. If the interpreter has been conscientious about portraying the intensely faulted situation that surely exists, then there will be main faults, others branching off from them, faults meeting each other, etc. On a drafted map these faults all look quite sure and definite. So wells are planned on the basis of the interpretation. Maybe two faults meet updip, isolating a nice wedge of sediments. "Let's drill here, in the updip corner." "How many acre-feet are there in the prospective field?" "What are the recoverable reserves?" What hogwash! Those faults may meet because of a guess! They had to be drawn in some directions, and they happened to meet there.

Of course, sometimes there is good evidence for the directions of faults. But usually the faults are visible on the sections, clearly or not so clearly, with no reliable way to map them on 2-D data. This type of problem is largely resolved by 3-D shooting. That alone is a powerful argument for using 3-D in any critical situation, as when wells are to be drilled on supposed fault traps.

On a time slice made from a 3-D seismic program, faults are seen as they would appear on maps. Reflections appear broken and shifted along the break. Most of the fairly straight alignments on time slices, especially the alignments that do not follow the reflections, are probably indications of faults.

Combining fault clues on slices and sections gives the best information on the faults. A slice shows the direction of the fault trace. A section shows the throw of the fault. A display that combines the two allows the fault to be interpreted most reliably.

Conformation

"How they were laid down"

Looking at the shapes of and interrelationships between reflections in groups, rather than single horizons, gives information on how the rocks were laid down. This is what is generally known as seismic stratigraphy, although it directly involves only deposition, not lithology, which is also part of stratigraphy.

Seismic Stratigraphy

Most of the improvements in seismic exploration are geophysical, that is, they are advances in the technology of geophysics, often improving the quality of sections. But seismic stratigraphy is a new type of advance, a geological one. Seismic sections have become such faithful reproductions of the subsurface the sections could be analyzed as though they were photographs of sheer cliff faces. A collection of such photographs could be studied to give information on the ways the layers were deposited—some delta deposition here, deep-sea deposits there, windblown deposition another place. The sections, like the photographs, don't tell the composition of the rocks, but do show the patterns they formed in being laid down.

A depositional sequence is a band of layers that were deposited more or less continuously. One such band is separated from the next by a period of non-deposition, and often erosion, an unconformity. A depositional sequence, as seen on a section, is called a seismic sequence, in a departure from the usual way of just calling the representation of a phenomenon on a section by the name of the phenomenon (for example, evidence of a salt dome on a section is called a salt dome, not a seismic salt dome). The reflections in one sequence may be parallel or nearly so. Those in the next sequence above or below might be tilted with respect to the first sequence, or form a different pattern.

Most sediments are laid down at sea, far enough out to be deposited fairly flat. This makes it usually safe to assume that the beds were flat before they were distorted by tectonics. So we flatten a seismic horizon and assume that this restores the section to the geologic time the horizon was deposited. However, at the edges of the seas, the deposition isn't necessarily flat. Material may have been washed down a slope in various layers as storms came or seasons changed or whole cycles of rainy or dry periods came along. The shapes the deposits formed in coming off the slopes can show fairly distinctively just how the deposition occurred.

Then the land rises or the sea level drops—it doesn't matter which, it's the relative change that matters—if the sea becomes lower with respect to the land, then the land that is exposed isn't likely to have deposition occur on it. It may be eroded. This is a halt in deposition, the end of a sequence of deposition. Then when the sea level rises again relative to the land, deposition can again take place. It is very likely that the new deposition won't be exactly parallel to the earlier layers. So this slope is the best place to see the break on a section. Each group conforms within itself but does not look like the other (Fig. 9-10). Now having determined these sequences at the slopes where they are fairly easy to recognize, the interpretation can move out to sea where

Fig. 9-10 Three seismic sequences
Credit: Petty-Ray Geophysical Division, Geosource Inc.

beds were mostly deposited flat. A break between sequences won't be recognizable there, so it must just be picked like any other reflection on out to sea. Picking all the unconformities within an area of shooting breaks the whole volume of data into the various sequences of deposition that occurred. Then looking at how the reflections slope, whether they form sort of an "S" curve or are just parallel or concave upward or something tells different things about the deposition.

Seismic stratigraphic interpretation is done by recognizing the seismic sequences and picking the unconformities between them (Fig. 9-11). Doing it thoroughly requires mapping all the unconformities over an area. This is considerably more effort than just picking from one to a few horizons for structural mapping. For practical interpretation purposes, with limited time available for interpreting, just determining the form of some of the deposition is often enough, without thoroughly mapping all unconformities. A thorough investigation of seismic stratigraphy can be made by a person with plenty of time available.

There are other, more subtle investigations that can be made. Seismic sequences can be plotted on a scale of geologic time, with gaps representing the periods of non-deposition. Changes of sea level through geologic time can be determined from a section and plotted. These plots can be correlated from place to place all over the world. This provides a way for geologic age to be determined from seismic data.

Constituents

"What they're made of"

There are several main ways that seismic exploration can yield information on the constituents of rocks and their pore-filling materials. Variations in reflection character, calibrated with well data; the seismic attributes of velocity, amplitude, frequency, polarity; seismic lithology, with amplitudes varying with angle of reflection—all these provide clues to the types of material in the subsurface.

Character Variation

Differences in rock types form traps that are nearly "pure" stratigraphic traps, from a seismic viewpoint. If a formation is porous downdip, and becomes impermeable updip, it can hold oil, even without any structural uplift. A map of the configuration of the horizon only shows the dip, necessary to the trapping, but no different from any other regional dip. So normal seismic mapping of a horizon does not deter-

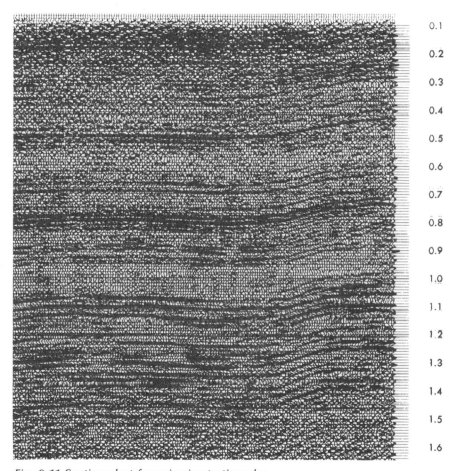

0.1
0.2
0.3
0.4
0.5
0.6
0.7
0.8
0.9
1.0
1.1
1.2
1.3
1.4
1.5
1.6

Fig. 9-11 Section shot for seismic stratigraphy
    Credit: Kim Tech, Inc. a subsidiary of Bolt Technology Corporation

mine that the trap is there. Much thought and effort has gone into detecting such traps, and in some special cases they can be found. One clue to the differences is in reflection character.

The character of a reflection, as seen on a section, is not a single wavelet, but a complex of reflections including those from any minor layering nearby. If there is a change in rock type or thickness, the character of the complex will change. This change of character is brought about by changes in strength of the wavelets because of velocity changes from the different types of rock, and different combinations of wavelets as thicknesses change.

If the characters could be identified and classified into groups like a limestone character, a sandy character, etc., that would be great. But they can't. What can be done is that the changing character of a reflection can be calibrated in an area that has many wells with seismic lines shot across them. Then it may be that, for one reflection in one area, the character changes correlate with lithologic or thickness changes in the wells. If so, then it can be possible to predict a rock type or bed thickness before a well is drilled.

A reflection may, for instance, consist of two strong peaks and some lesser wiggles. But in places that character may change to one strong peak with a dent in its top and a strong trough. If that doublet character appears in the vicinity of a well in which the formation has a porous sand, then an interpreter may look at the character of the reflection near other wells. If the doublet appears to be consistently associated with porosity, it becomes an exploration tool, and wells may be recommended on the basis of doublet character of that reflection.

Seismic Attributes

Some seismic investigations that are made to aid in configuration exploration have turned out to yield information on the constituents of the rocks. Velocity determinations, necessary for stacking and for migrating, and conversions to depth—to find the highs and lows—also give clues to lithology.

Analyses of frequency, polarity, and amplitude also yield clues to makeup of the rocks. These and velocity are called seismic attributes. One or another of these is often displayed as an overlay of color on a normal configuration section.

Amplitude Anomaly

An amplitude anomaly is a part of a reflection with an unusually high or low amplitude for that reflection.

A bright spot is a short length of a reflection that is considerably stronger than the reflection normally is (Fig. 9-12). Its importance is that it may indicate hydrocarbons. A seismic reflection occurs where there is a velocity interface, that is, the sound is traveling through rock at one velocity, and then encounters rock that transmits sound at a different velocity. The difference in velocity causes the echo, or reflection, and the greater the difference the stronger the reflection. Also polarity is affected. A change from slower to faster causes the trace to swing one way, while a change from faster to slower makes it swing the other way.

So a reflection from the interface between two rock layers should look about the same from place to place, as long as the composition and thickness of the layers remain the same. But porous rocks have fluids in them—water, gas, oil, even air if nothing else. And these fluids

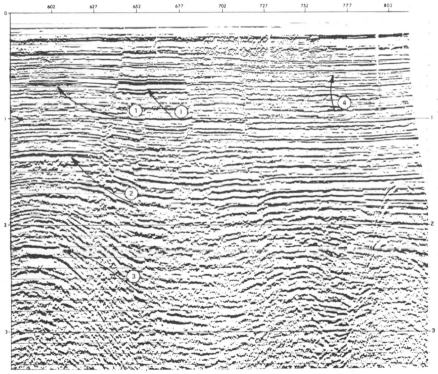

1. Bright spots — gas-noncommercial (too thin)
2. Bright spot — gas-commercial
3. Flat spot — condensate-commercial
4. Bright spot — Lithologic change, not gas

Fig. 9-12 Bright spots and flat spot   *Credit: Seiscom Delta United*

transmit sound at different speeds. So the sound velocity of rock is influenced by the fluid in it.

Now consider a porous rock layer with the pore spaces filled with the usual subsurface fluid, salt water (Fig. 9-13). The water in the pores has a velocity of about 5000 ft/sec, while the rest of that rock has a higher velocity, maybe 20,000 ft/sec. The combination of rock and water has a velocity somewhere between the two, depending on the proportions of rock and water, maybe 10,000 ft/sec.

Velocities generally increase with depth, so let's assume the layer above it has a velocity of 8000 ft/sec. Then the strength of the reflection is a result of that 2000 ft/sec difference between the 8000 and 10,000.

Also the change is from a lower to a higher velocity, so the polarity is called positive, with the trace breaking, say, to the right. Now if, for some reason, some extent of the porous rock contains not water, but gas, the velocity relationship there is not the same. The gas may have a velocity of something like the velocity of sound in air, 1000 ft/sec, or, as it is compressed underground, somewhat higher.

The velocity of the gas and the rock's own 20,000 may combine to make, say, 5000 ft/sec. Now the difference at the interface has changed from 2000 ft/sec to 3000 ft/sec, so the reflection should be stronger. Also, the change is now from a higher to a lower velocity, so the polarity is changed from positive to negative.

So the section has a reflection that, for a limited distance, is stronger than normal and has its polarity reversed, so what is normally a peak becomes a trough. Then on the other side of that distance, it changes back to normal again. That is the most obvious kind of bright spot to look for—a short length of a reflection that is outstandingly strong. The polarity may not be reversed, and if it is, the reversal may not be easy to

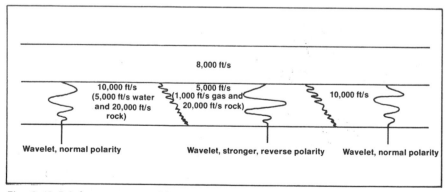

Fig. 9-13 Bright spot caused by presence of gas in rock

detect, as it may seem that the reflection moves up or down a little instead.

With various velocity arrangements all sorts of things can occur. The velocity contrast can be less than normal instead of greater, so the exceptional part of the reflection is unusually weak. This makes, not a bright spot, but a dim spot. The velocities can change, but remain in the normal direction for the reflection, so the polarity is the same rather than reversed. A trouble with a dim spot is that, although it may be just as significant as a bright spot, it may not show up as well. And there are other reasons for reflections to be weak—near-surface variations, poor geophone plants, etc. Bright and dim spots are spoken of together as amplitude anomalies.

A dim spot is particularly likely to occur in a reflection at the top of a limestone, as the lime has a fairly high velocity. A limestone layer or a reef may thus contrast greatly in velocity with the overlying material, maybe a shale, and produce a strong reflection between the two. If the upper part of the limestone contains gas, its velocity is reduced. That may make the velocities of the lime and the shale more nearly the same. So the reflection is poorer there, a dim spot. It may have the same polarity as the normal reflection, or if the gas pushed the velocity below that of the shale, the polarity would be reversed.

Amplitude anomalies are best displayed on true amplitude sections (Fig. 9-14). On those sections, the reflections are displayed in truly relative strengths, rather than being adjusted toward equalizing ampli-

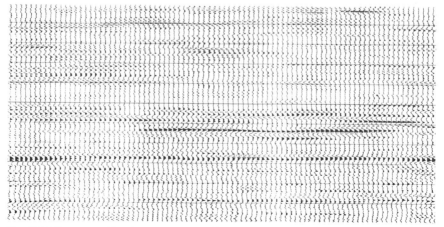

Fig. 9-14 True amplitude section Credit: Western Geophysical Co. of America

tudes. A true amplitude section shows most reflections looking rather weak, so the strong ones stand out. It is easy to recognize the reflections that are strong all along their length, and also the ones that are locally strong, in bright spots. Dim spots are also in their relative relationships to other parts of the reflections, but do not stand out as clearly in a glance at the section.

There is an aspect of bright spots in normal AGC sections that makes them about as recognizable as on true amplitude sections. If there is a strong bright spot in the data, then, in the processing to equalize the reflections, the bright spot is toned down to more nearly the amplitude of the other reflections, and of the same reflection in other places. But just one reflection isn't toned down alone. After all, if the equalizing was perfect, there wouldn't be any reflections left on the section at all, just dead traces. In reducing the amplitude of the strong reflection, the reflections above and below it are also reduced to below normal amplitude. The attempt to equalize leaves the bright spot still somewhat stronger than normal, and the adjacent reflections weaker. This leaves a shadow zone of dimness around a bright spot. This shadow is quite easy to recognize, and is often as good a way to recognize a bright spot as looking at a true amplitude section. But that is only recognizing, not measuring. A true amplitude section is more numerically correct, and shows better just how bright the bright spot is.

But in detecting possible gas in formations by bright or dim spots, the anomalous amplitude does not give a good idea of how much gas there may be. Not much gas is necessary to change the velocity of a rock, so an amplitude anomaly may indicate very little gas, maybe as little as 2% of the volume of the rock. The bright or dim spot indicates that there may be gas, in a more direct way than seismograph does otherwise, but doesn't tell much about whether it is in commercial quantities or not. And lithologic variations, coal beds, etc., may produce bright or dim spots in the same way gas does.

This discussion has been primarily about gas, as it has a much slower velocity than water. Oil has a velocity only a little slower than water, so it doesn't make as big a change in the velocity of the rock.

Careful analyses of amplitude anomalies are made, calculating coefficients of reflection at the various velocity interfaces, to put relative amplitudes on a more quantitative basis.

Flat Spot

A phenomenon similar to a bright spot is a flat spot, a bit of a reflection that is relatively flat in relation to its surroundings (refer back to Fig. 9-12).

What makes for a flat reflection? A flat velocity interface. Then what makes a flat velocity interface? A flat-lying rock layer would, of course, but it would probably not be unusual for the vicinity, and so wouldn't stand out as having special significance. But there is one thing that is almost always pretty flat, the top surface of a liquid. And if the liquid has a gas above it, rather than another liquid, there is a large difference in velocities at the surface. If there is a gas-water or gas-oil contact, the top of the water or oil should be flat. If all the reflections around have some dip, except for one flat reflection among them, then it may well be that the flat reflection is from the velocity difference between a gas and a liquid.

A oil-water contact too is flat, and may even appear on the section, but the velocity contrast is much less, so it is not likely to be notice-able.

The ideal situation is to see a structure on a section, with a flat reflection in the top of it. The structure may contain gas down to that reflection.

A number of flat spots on one or both sides of a fault may indicate a seepage of gas up the fault plane, displacing water in porous beds along the way.

Some factors can make a "flat" spot not exactly flat. A lateral change of velocity in the rocks above may give a tilt to the reflection on the section because of the velocity difference. "Flat" spots aren't necessar-ily flat, or even straight and tilted. The slow velocity of the gas above the contact will tend to make reflections below it appear lower than they ordinarily would, in velocity pulldown. If there is a flat gas-liquid contact, and the upper surface of the trap, the seal, is arched, as in a structure or a reef, then the pulldown will be greater in the middle than at the edges, so the "flat" spot will sag downward in a bowl shape. If the trap is wedge shaped, as in a pinchout, the "flat" spot may be straight but angled downward. Or there may be a tilted contact in the subsurface, caused by hydrodynamic forces in the slow migration of fluids in the porous rock, that allows the liquid surface to be slightly tilted.

There may be more indirect relationships between a liquid surface and a flat reflection, as when cementation has taken place by a mineral coming out of solution in the water. But above water, the oil or gas would not have held the mineral in solution, so there would not be cementation there. This lithologic difference can cause a flat spot at an oil-water or gas-water interface. This type of flat spot can indicate that hydrocarbons are trapped there—or were once trapped there. It can be tilted for the same reasons as apply to other flat spots, and for one additional reason. If the bedding has been tilted after the cementing material was deposited, the velocity contrast is likely to be preserved. In this situation, the oil or gas may still be in the tilted trap.

So the thing to look for is a reflection that is flatter than the other reflections around it, preferably with sharply defined ends, a reflection that does not look like it could be showing the attitude of rock layering, or one that is turned downward. It should also not look like it could be a multiple of a shallower horizon. Then it can represent the upper surface of a liquid, the lower surface of a body of gas. And where there is gas, there may be oil beneath it. The gas, though, does not have to be composed of hydrocarbons. It can be carbon dioxide or any of the exotic gases sometimes found in the earth.

Seismic Lithology

Seismic lithology is another means of getting information on the constituents of rocks from seismic data. In contrast to the geological use of the data for seismic stratigraphy, this method is highly seismic. In seismic lithology the traces from one reflection point are not stacked, but are considered separately. There are a number of traces that all give information about the same, or nearly the same, point in the subsurface. But they are derived from different offsets, different source-to-receiver distances, so they were reflected at different angles. It happens that these traces from different angles do not show the same reflection amplitude or character. That is, sound reflects better at some angles than at others. The differences in amplitude depend on the makeup of the reflecting surface—the type of rock, the lithology, of that surface. The indications are not definitive, they don't allow a person to state, from that information alone, "This reflection is from a sandstone". But they may make it possible to state with confidence, "That reflection is not from a shale". The content of the pores in a rock also has differing effects at different offsets. So a person might be able to say, "This reflection is likely to be from a sandstone with salt water in it". That can be useful information. And combined with other information, like nearby well data, it can be quite definite. This method is used for specific localities, like well-developed prospects. The name is a little misleading, as it is only one of several techniques for investigating lithology by seismic means. Another name for it is offset dependence, which refers more to the technique than the hoped-for result.

The form the data takes is an NMO-corrected CDP gather (Fig. 9-15). The traces (that are combined into a single trace by stacking) are for this purpose played out side by side. With NMO correction, the reflections are flat. The differences in amplitude of the various traces can be

OFFSET ⟶

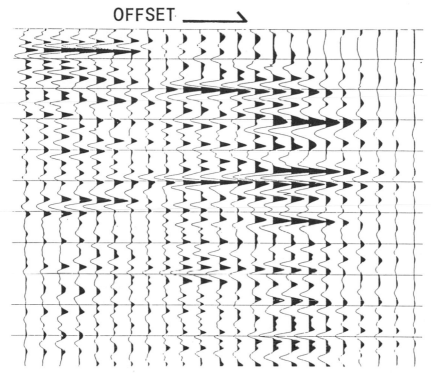

Fig. 9-15 Amplitude variation with offset   *Credit: Terra Linda Group, Inc.*

readily observed. Some lithologies have the greatest amplitude on near traces, some lithologies are strongest at greater offset, others do not have much amplitude difference with offset.

There are other factors besides lithology that complicate the interpretation. Considerable calculation is done to correct for the other effects. The result is plotted as a graph of amplitude vs. offset. Then tieing that information in with well data and other information gives an interpretation of the lithologies of the reflections.

Seismic lithology analyses are performed on the traces from a few shots to provide information on one restricted vicinity. They are not needed at every shot along a line, but at places where wells are planned, or to establish some lithological identifications that can be used in the normal picking of sections.

CHAPTER 10

Other Seismic Methods

Besides conventional shooting using reflection of sound waves, there are other types of seismic exploration. One uses reflection of shear waves. Another uses refraction of sound (compressional) waves.

Shear Waves

There are several kinds of waves that travel through the earth. Compressional waves are sound waves, the waves normally used in seismic exploration, the kind that have been discussed so far in this book. In compressional waves, the wave consists of alternating compressions and decompressions. The particles of the rock they go through vibrate forward and back. Shear waves are transverse waves in which the particles vibrate from side to side. They are not sound waves, as they are not compressional. But they are also vibrations, and are reflected and refracted like any waves.

Compressional waves are also called P-waves or primary waves, and shear waves are called S-waves. The two kinds of waves not only travel through the earth in different manners, but they have different velocities. Seismic velocities for S-waves are about half the velocities for P-waves. The P and S come from the Latin words for "first" (primus) and "second" (secundus), because the higher-velocity P-waves are the first waves seen on an earthquake seismograph, and the S-waves come in second.

One way to initiate compressional waves in the earth is to hit the ground with a downward hammer blow. Shear waves would be initiated by hitting a stake in the ground from the side with the hammer. In seismic field work on land, S-waves are generated with a vibrator that presses its weight against the ground and imparts a side-to-side vibration to it. Some vibrator trucks are equipped with both types—an up-and-down vibrator and a side-to-side vibrator. Or, instead of a vibrator, a truck-mounted hammer may strike the side of a weight that is pressed against the ground under the truck.

To receive shear waves, a geophone must be sensitive to sideways motion. This is achieved by making a geophone with its coil and magnet suspended horizontally, rather than vertically as in a conventional P-wave geophone. The sideways motion of shear waves can be in any direction around the direction of propagation of the wave. Of all these directions, two are named: the horizontal SH waves and the vertical SV waves. To record all three—P, SH, and SV—a three-component geophone has elements free to move in three directions at right angles to each other.

P-waves and S-waves produce similar-looking sections. But to make the sections easy to compare, compensation is made for the different velocities by printing the S-wave section at half the time scale of the P-wave section (Fig. 10-1). In most situations a P-wave section has better reflection quality than an S-wave section. This is to be expected, as the industry has devoted many years of effort to the improvement of P-wave data. But in some situations, S-wave sections have better reflections. The two kinds of waves respond differently to some features of

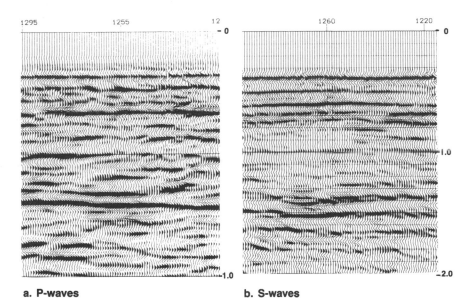

a. P-waves   **b. S-waves**

Fig. 10-1 Scale comparison
         Credit: Petty-Ray Geophysical Division, Geosource Inc.

the subsurface, so if both are available, more information can be obtained than from either one alone (Fig. 10-2).

Any feature that looks slightly different on the two can be analyzed better by comparing the two. The main difficulty in making the comparison is in the different velocities of the two kinds of waves. But with the shear wave section at half the vertical scale of the P-wave section, reflections from one formation are fairly near the same place on the two sections. Then the interpreter looks for a distinctive reflection or dip on a reflection, that is evidently from the same subsurface feature on both sections. With that starting point, other features of the two sections can be compared.

The ratio of P-wave velocity to S-wave velocity varies somewhat for different kinds of rock, and this can give clues to rock type. The depth to the formation is the same, no matter what the velocity, so the reflection times are in the same proportion as the velocities. Therefore, if a reflection can be confidently identified as the same one on a P-wave section and an S-wave section shot in the same place, the ratio of the times can be used for information on the lithology.

A major difference in the two types of waves is in the way they behave in fluids, both gases and liquids. P-waves travel through gases at slower velocities than through rock, so a gas-filled rock has a considerably slower velocity than the same rock with the pores filled with a cementing material. But S-waves aren't propagated through fluids at all. They travel through the rock, bypassing the pores, and are only slightly affected by the fluids. P-waves are alternate compressions and rarefactions of matter, affected by the compressibility of the matter. But shear waves, in not being affected by fluids, tell things about the rocks themselves.

A bright spot is caused by a locally greater than normal velocity contrast between two layers. This is usually a result of a slow velocity in part of a layer. The slow velocity can be a result of either gas in the formation, or a local difference in the type of rock. A bright spot produced by gas will appear on a P-wave section, but not on an S-wave section. But a bright spot created by a change in the type of rock will be on both kinds of section. This difference is a way of distinguishing bright spots that indicate hydrocarbons from bright spots that do not.

Similarly, flat spots that are reflections from upper surfaces of liquids overlain by gas will appear only on P-wave sections. If a flat spot also appears on an S-wave section, then it is caused by some flat rock surface. It might be a multiple of a flat bed above where it appears to be. Or it could be a one-time liquid surface that has been cemented in one fluid (water) but not in the other (oil or gas).

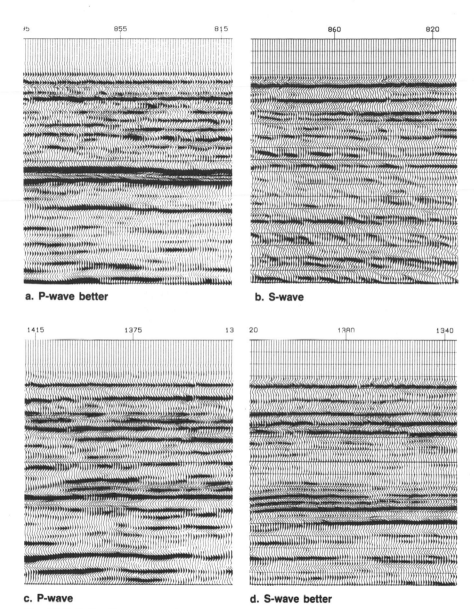

a. P-wave better

b. S-wave

c. P-wave

d. S-wave better

Fig. 10-2 Reflection comparison
         Credit: Petty-Ray Geophysical Division, Geosource Inc.

When compressional wave energy goes straight downward into the earth, at each interface it vibrates the rock and sends out new P-wave energy in all directions. Naturally. A compression hits the interface like a hammer blow, and sets it to vibrating. But when it hits the interface at an angle, its compressional force hits an interface somewhere between downward and sideward. The sideward component, like a sideward hammer blow, sets up shear waves. So the compressional wave generates both kinds of wave in the rock. There is a critical angle beyond which most of the reflected energy received at the top of the ground is shear-wave energy. That is recorded on an S-wave geophone group far enough to receive it from the critical angle. A shear wave striking a surface at an angle also generates both kinds of energy. The energy of one type that was generated by the other type is called converted energy.

In water, even though S-waves do not travel through fluids, S-wave data can be recorded by way of converted energy. P-waves go into the water at an angle and at the sea bed generate S-waves, which travel downward and are reflected back up. Arriving again at the sea bed, some of the energy is converted back to P-waves. So P-waves can be initiated in the normal way and recorded in the normal way, and S-wave information obtained from them (Fig. 10-3). This information is distinguished from altogether P-wave data by its wide angle energy path (so it is received at far geophones) and by its greater NMO (because of the slow shear wave velocity). And it is distinguished from refracted P-waves, which also are received from large offset distances, by its curving NMO pattern as opposed to the straight-line pattern of refracted energy.

On land, the use of shear waves in seismic exploration was introduced a few years ago. Its use, although still a small proportion of the shooting done, is increasing. Offshore, it is just beginning to be used. Also, some VSPs are being recorded with three-component geophones to simultaneously record P-, SH-, and SV-waves.

Refraction Exploration

Refraction of sound plays a large part in seismic exploration. It bends the paths of energy in reflections. It yields data for near-surface corrections of reflection sections. And there is the refraction method of exploration, as distinguished from the more common reflection method.

The principle of refraction is that energy is bent, in going from one medium to another that transmits it at a different velocity. Light, sound, radio waves act this way. The best-known example of refraction is the

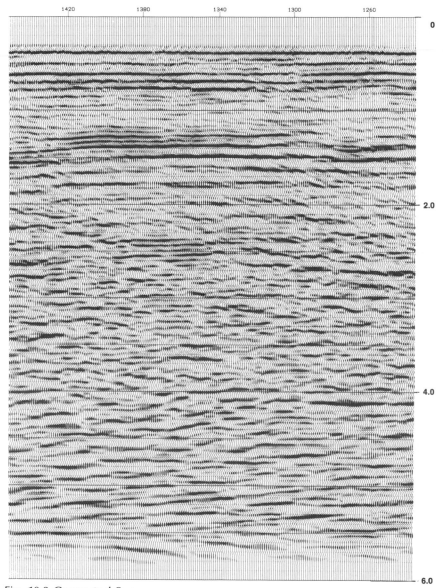

Fig. 10-3 Converted S-waves
Credit: Petty-Ray Geophysical Division, Geosource Inc.

stick poked into water, and appearing bent at the surface of the water. Light travels at different speeds in air and water. There is also reflection at the surface, so sometimes you see in the water both the refracted image of the lower part of the stick and the reflection of the upper part. The reflection you see is the light that went from air to water and bounced back, while the refraction is the light that was bent in going from the water into the air.

Seismic energy traveling down into the earth is similarly bent at velocity interfaces that it does not strike perpendicularly. There is a definite amount of bend for each velocity contrast and angle of approach to the interface. When there are a great many interfaces, the energy path, making all those little bends, becomes nearly a smooth curve. When the velocity increases steadily, as in shale that is compacted with depth, the bending does become a smooth curve.

The uses of refraction to gather information for near-surface corrections and for refraction surveys depend on having the receivers considerably farther from the source than the depth of the horizon.

From a seismic source, sound goes out in all directions. Every bit that strikes an interface sets it to vibrating, as though there was a new sound source there. Then from each of those points, sound goes out in all directions. The part that goes upward is seismically useful as it can be received on the surface. In general the sound that gives the useful information to a geophone is the part that took the quickest path from source to horizon to receiver. If the instrument is near the source, then the sound to reach it by the quickest route from the horizon is that which was reflected from a point midway between source and receiver. However, if the receiver is far from the source relative to the horizon, and if, as is usually the case, the lower layer of the interface has the higher velocity, a quicker path will be down at an angle to the faster layer, along it, and back at an angle to the surface. This is like the decision you might make when walking across a ploughed field. If there is a road nearby, parallel to your course, then, depending on the distance you are going and how far away the road is, it may be quicker to detour over to the road, and walk part of the way along it, than it is to walk straight across the field.

Look what happens when the refraction path is quickest. There are groups of geophones spread out in a line. Each group receives energy that has gone down to a fast layer, along it, and back up (Fig. 10-4). For all the groups, the route down is the same one, and the paths up to the surface are about equal to each other. Topography may make them different, but that can be corrected for. After correcting, the up and down paths to a group are about the same, but there is a difference

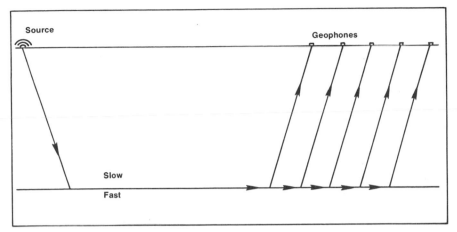

Fig. 10-4 Sound refracted along fast layer

from group to group in the time the energy spent in traveling through the faster layer. The distances between the groups are known, and the differences in the times recorded at those groups are the travel times for those distances through the high-velocity layer. Divide distance between groups by time difference between traces to get velocity. And notice that this difference is a one-way time; it doesn't need to be divided by two like the two-way time in reflection calculations. If the times are plotted on a graph against horizontal distance from source to geophone, a slanting line results (Fig. 10-5). As time in the fast layer is the thing that was measured, then the slope of the line represents how fast the sound went through that layer. The intercept—the time reading of the slanting line at zero horizontal distance—represents the time the energy took going down and up. So the slope is the velocity in the fast layer, and the intercept is the time spent in the slower zone.

Those principles apply only if the horizon is flat. If it has dip, then the dip will also affect the slope of the plotted line. To get around this problem, a refraction line is also shot the other way around, and the slopes of the two plotted lines averaged.

This kind of plot is the basis of both uses of refraction. For near-surface corrections, the regular geophone spread for reflection shooting is spaced for obtaining reflections from the depths of interest. But the base of the LVL is so shallow that, for it, most or all of that same spread is properly spaced for receiving refracted energy. So this near-surface refraction gives the velocity of the LVL, and aids in determining depth to the base of the LVL (Fig. 10-6). The refracted energy in this case is the first energy to arrive at the farther traces, and it makes the traces

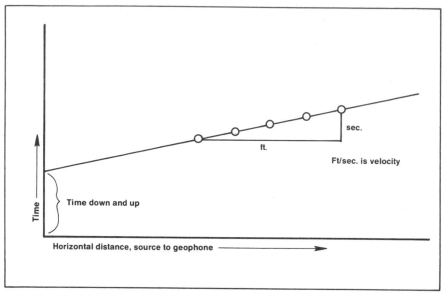

Fig. 10-5 Refraction plot

"break", deviate from fairly straight paths. These deviations are the first breaks.

When surface sources are used in reflection shooting, a special small crew, an LVL crew, may be used to obtain information on the thickness and velocity of the LVL by refraction, to supplement the data from the surface source. For distinct refraction first breaks, each geophone group consists of only one geophone. The LVL crew may drill shallow holes and fire small charges in them, or it may use an air gun mounted on a small vehicle (Fig. 10-7). Or the regular crew can record some data especially for the first breaks. One device allows a crew to leave the normal reflection spread on the ground, and use a switch to record from just one of the geophones in each group.

The very same geometry, but on a larger scale, is used in refraction exploration. In order to have the instruments far away in proportion to the depths of mapping horizons, they are placed several miles from the shot point. A refraction crew, because of this greater distance, has its own special problems of logistics. More vehicles may be needed than on a reflection crew. Communication between units is normally by radio.

From one to several horizons will be mapped from the refraction data. The velocities of the formations themselves—not average velocities of everything above them as in velocities from normal moveout—

are obtained. This is one of the major advantages of refraction shooting. There is little trouble with continuous identification of a horizon. After a gap in data or a fault, the refraction with the same or nearly the same slope is the same horizon. Its big disadvantage is that the depth determinations are much less accurate than those obtained from reflection shooting. The poor depth information makes people use refraction as the main type of investigation, only when reflection will not work for some reason.

Refraction was the first type of seismic exploration to be in common use. Salt domes were found by refraction before reflection became

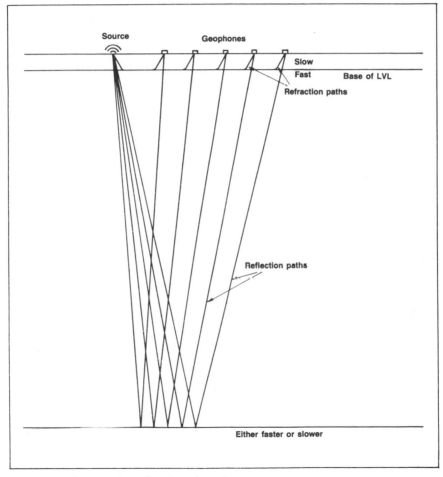

Fig. 10-6 Refraction in reflection shooting

Fig. 10-7 Air gun for LVL crew   *Credit: Bolt Technology Corporation*

popular. The high velocity of sound in the salt was detected on refraction lines that radiated from a point in a fan arrangement, therefore called fan shooting. A quicker energy reception on one line than on another indicated that the first might have gone through a salt dome.

Refraction is now sometimes used in mountainous regions where terrain is too rough to allow occupying all the positions needed for a reflection survey. It is also used in areas that have not been explored, to give velocity information and, from it, rough identifications.

There is a simple means of obtaining refraction data economically in the course of an offshore reflection survey. During the reflection shooting, the boat runs along straight lines, with geophones trailed on a cable, and shots (air gun, gas gun, etc.) fired at frequent intervals.

The description of refraction shooting given in the preceding segment utilized a single source and a spread of geophones. But source and receiver are interchangeable, as far as energy paths are concerned. So one geophone group and a number of shots give the same information as one shot and a number of geophone groups.

Taking advantage of this interchangeability and of the shots that are being fired anyway, a sonobuoy is used. It is a device consisting of one or more geophones built into a plastic tube about three feet long, that floats upright and contains a radio transmitter. The sonobuoy is thrown overboard. A battery is activated either immediately before it is thrown, or by the contact with sea water. The geophones are released, to hang suspended on a wire, maybe 100 feet down in the water. The geophones receive energy from the shots fired for reflection. The radio broadcasts the output of the geophones. The broadcast is received and recorded on magnetic tape on the boat. As the shots continue and the boat gets farther away, refraction data is recorded. The sonobuoy is not retrieved, but is designed to sink in a few hours.

So in this case the refraction information is a sort of byproduct of the reflection survey. It also has some of the disadvantages of a byproduct. The sonobuoy is not in a fixed position, but may drift with wind and current. The line is shot in only one direction, not reversed. The shots are normal strength for reflection that is to be stacked, and may not provide as much energy as would be best for refraction. But within these limitations, quite useful information can be obtained, to be tied in with the reflection data. The fact that the two are shot together, with the same surveying, the same tide, the same currents, can be an advantage in working them together.

A variation of the sonobuoy is another telemetry device for offshore refraction shooting, also a floating tube with geophones and radio transmitter, but re-usable and employing an anchor. Several may be anchored along a line. Then the boat can shoot past them and on beyond in a refraction line. This method gives refraction data that is shot as such, not as a byproduct of a reflection line. Surveying, distances, strengths of shot can be tailored to the purpose. And the tubes with their geophones do not drift, but stay in more or less the same place. Changing wind and current directions may cause them to swing from the anchors in different directions, and strong currents can cause them to drag their anchors, but these changes are usually not great during the shooting of the line.

Other Uses

Refraction seismic surveys can be made fairly easily to investigate shallow horizons. A weak source, a short spread, and instrumentation only sensitive enough to record first breaks are adequate to investigate down a few tens of feet into the earth. The source may be a sledge hammer blow on a metal plate on the ground. Not many traces are needed, maybe six to twelve. And first breaks are best recorded by single geophones rather than groups. So the equipment can be very portable.

This type of refraction survey is useful in archaeology, along with magnetometer and ground-penetrating radar, to find buried remains of walls and the like. It is also useful in engineering work, for instance to determine depth to bedrock before building construction is begun. In mining, refraction may be used in conjunction with radar. In all these uses the location surveying is simplified by the fact that only short distances are involved.

CHAPTER 11

Old Data and New

Improvements are constantly being made in seismic techniques. The best shooting and processing are usually done with the most modern techniques. But there is also a lot of old data around, and a person dealing with geophysics needs to know what to do with some of the older forms that may be encountered from time to time. Sometimes very useful information can be obtained quickly and economically from old data. And then there is the problem of what new data should be saved and how it should be stored.

Availability of Old Data

Shooting is normally done by a contract crew working for a client, and the data belongs exclusively to that client; or it is done by a company crew, so it belongs to the company. Seismic data is usually highly confidential. This secrecy has been broken down a bit by the idea of trading data. Companies are willing to let out data from areas in which they no longer have an interest in exchange for some in an area they'd like to know about. The trading is cumbersome, and takes up a lot of employee time that could better be spent hunting oil instead of digging out old data and getting agreements signed, etc.

So the trade service came along, an organization to handle the legwork of trades for companies, and sometimes store the data so the company wouldn't have to hunt for it. From that aspect of the trade service, the geophysical library developed, to distribute maps of data available from different companies and handle the trades for the oil companies. Joining the library required depositing some set amount of data, with that deposit allowing the company to take out from two to four times that amount of data from other companies' contributions.

The library was fine among large companies, but smaller companies without a backlog of old data were pretty well left out. And often they were the very companies that would find the old data most useful. So libraries shifted emphasis from trade to sale of data, and the advantages of money over barter were discovered anew. A company can deposit any amount of old data it would like to earn some money from, and any company can purchase data without having to have data on deposit. The expressions data brokerage and data exchange are now used as well as trade service and library to describe the companies that handle sales and trades of data.

Companies still trade data individually, but the sale of data through a data brokerage is the most common way of handling the transfer of seismic data between companies.

Much of the old shooting is available for trade or purchase. The brokers' maps of available data make it easy to determine which control covers the area of interest. And the maps simplify putting together a combination of parts of different companies' shooting to form a survey giving as good coverage over the area as possible. Some maps of available data are stored on magnetic tape or disk so the search can be made by computer.

Trade or purchase of old data usually gives a company a copy of the sections and the right to copy the tapes. But it is usually stipulated that the data is for that company's use only, not for any other company. However, this is customarily interpreted to mean that data can be shown to partners in drilling deals, etc.

Price of old data varies with the type of data and the company offering it. But it is usually much cheaper than acquiring new data.

Value of Old Data

A basic question about old data is whether there is any point in bothering with it at all. Didn't the company that shot it find all the anomalies in it, and drill the good ones? For that matter, is there any reason to even look for oil in an area that somebody else has already shot? Are you that much smarter than all those other exploration people? As for looking in an area that was already shot, new fields are constantly being found where people have been looking for a long time.

On using the old data, you don't have to be smarter to get something useful from the old data. First, it gives you regional information. Then, your outlook isn't the same as the other company's was. The leases you are considering may not be the same as theirs, you may be interested in a different formation, or you may be looking for smaller features or a different kind of feature. Putting together parts of two or more old surveys gives coverage that none of the companies had after shooting their own prospects. And, last, maybe you can be smarter, or anyway think up a new idea about the area. Did anyone ever refuse to look at an electric log because other geologists had already worked with that log?

Old data can also be reprocessed to yield better information than before. The most striking example of this is in transcribing old single-shot paper records onto magnetic tape, so they can be processed digitally into seismic sections.

A remarkable change can take place in going from individual records to sections, and often a much better interpretation can be made. A section is almost essential for display to management or to other companies.

Some reasons for using old data are:

It is non-seasonal. You don't have to wait for a freezeup, or a dry season, to acquire it.

It is really turnkey. You know exactly what it will cost.

Sometimes it is the only kind of seismic data that can be acquired in an area. If shooting is no longer allowed for some reason, the old data can still be obtained.

In most cases, the old shooting isn't expected to determine a well location. The lines probably weren't shot just where you need them, and any hitherto-unnoticed lead must be pursued with the lines it indicates. So there will normally, unless the old work condemns the prospect, be some detail shooting to be done, or even a repeat of a key line or so to verify the prospect with modern techniques.

But the old pattern is often about as good for reconnaissance as the one you'd lay out at the start, and new detail lines will go where you want them to anyway.

Forms of Old Data

Advances in geophysics cause not only the quality but also the form of data to change. So old data, either from your company's files or obtained by trade or purchase, may be in any of several out-of-date forms.

For many years seismic data was recorded on individual paper records, like modern monitor records. The number of traces on a paper record increased through the years from two or three to 6, 12, 24, and 48 (Fig. 11-1). The records were made photographically, so they had to be developed in the dark. On a crew with a recording truck, this would be done in the doghouse, a little darkroom in the truck. Some early records were dark with white traces and timing lines; others were white with black lines.

The paper record was the only recording of the shot. There was no way of changing it after it was shot, so filters, gain, etc. had to be set before a shot was fired. Mixing was used. It was a means of making poor reflections more pickable by blending the information on one

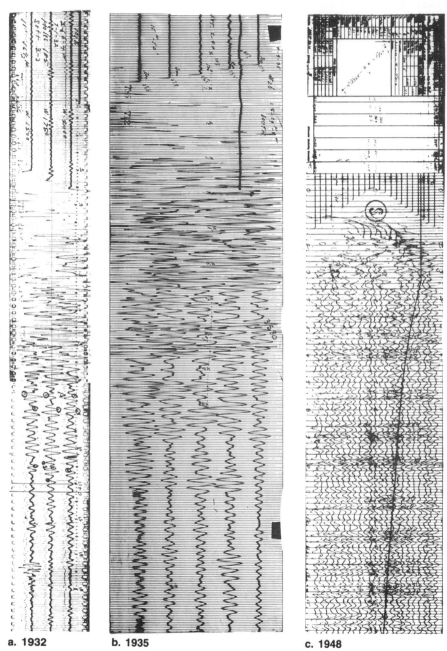

a. 1932   **b. 1935**   **c. 1948**

Fig. 11-1 Paper records (1948 record, credit: GTS Corporation)

trace with some from others. This smoothed the reflections some, but was misleading where there was dip in the subsurface or irregular ground.

The reflections curved with normal moveout and were distorted by topography and LVL. Corrections for all of these were made as calculations of milliseconds to add to or subtract from reflection times. Structure could be determined from the records but little or nothing about subtle characteristics of seismic data that are visible on modern sections.

Many of the paper records have been microfilmed. Some have been transcribed onto tape by people running sensors along traces or by computer. The transcribed data can be put into section form (Fig. 11-2). The filtering and mixing cannot be removed, but other processing steps can be taken. Traces can be moved up or down. Normal moveout can be removed. The data can be migrated. And the section will not have the stains and dark spots from poor developing that many of the original records had.

When tape recording entered the seismic process, analog tapes rapidly replaced paper records as the recording medium, and seismic sections were made from the tape. Their form was much like that of today's sections. But the early ones were only 100% sections, so they were not as good as CDP stacked sections.

Each shot was recorded on a separate piece of tape. The tapes were from 4 to 10 inches wide and 2 to 3 feet long—rather like the dimensions of the paper records they replaced. The tapes were kept flat in flat boxes, or rolled up in little plastic tubes. A tape was put on a drum in the recording equipment; the shot was fired and recorded; the tape was removed, annotated with shot point number, date, etc., and refiled. The tapes were similar to records also in having all the traces side by side. Twenty-four trace recording called for twenty-four channels on the tape.

If the analog tapes have since been converted to digital form, they will be on regular digital reels of tape. This gives some of the advantages of digital recording, but not all. The data is still limited by having been recorded in analog form. The frequency and amplitude data aren't as good, for instance. But with the data in digital form, digital bandpass filtering, deconvolution, migration, etc. can be applied to it.

Common depth point techniques came along during the time analog tape was in use. Improved processing can be applied to CDP analog tapes.

When digital tape began to be used, the forms of both sections and tape became essentially what they are today. But there have been numerous advances all along, so rapidly that sections made three years

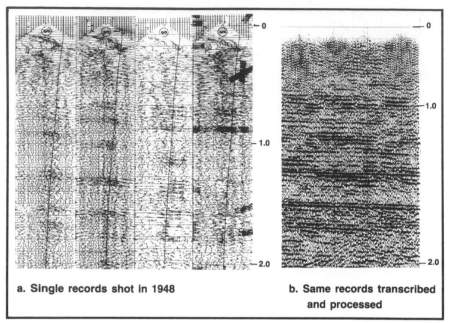

a. **Single records shot in 1948**   b. **Same records transcribed and processed**

Fig. 11-2 Transcription of old data   *Credit: GTS Corporation*

ago may be considered old data, and not as good as today's. In some respects, even last year's shooting is dated. Some of the advances are in field recording and some in data processing. The recording is already done and can't be changed, but the processing advances are largely available for the old data by reprocessing.

Combining Old and New

Once, a new boss of mine who had been out of seismic exploration for some years asked me to tell him the new developments. I was able to tell him of only one change that he hadn't heard of, and it was minor. That is hardly the case nowadays. The changes started with the introduction of seismic sections, and shortly thereafter, magnetic tape recording. Since then, the rate of change has been both rapid and accelerating. There is a tendency to feel that only new data has value, and that old data is so much poorer that it should not be used.

The advances are real, and the philosophy of using only new data is, in part, valid. It is often so much better that for some purposes old data just is not a substitute for new. In areas with severe multiple problems, in amplitude investigations, or for some seismic stratigraphy uses, the old data may not contribute useful information, and may have to be reshot in critical locations.

However, as a matter of personal philosophy, I feel that seismic data is expensive and hard to get, and should be used whenever some information can be obtained from it. In general, rather than reshoot an old line, I would shoot a new line offset from the old one far enough for both to contribute data to a map. A line through a well, or through the location of a planned well, should be the best available. If there is an old line through that location, then maybe a new line can go through that same spot, but in a different direction. One situation in which a line should be reshot, though, is when it is intended as a check on the old one, to see to what degree the old data can be believed. Then the duplication of one line may serve to calibrate a number of old lines, so the information they do have can be incorporated with the new data.

The above sounds like new data is superior just because it is new. We all know that, in general living, newer may not at all mean better. Well, in seismic data too, newer does not necessarily mean better. But improvements in equipment and techniques are so rapid that new almost always does mean better, and a good deal better, at that.

Back to philosophy on the use of old data. The natural, minimal effort, way to work is to map the lines as they are shot, so the first shooting in an area forms a framework for the interpretation. As new lines are added, they are fitted into the framework, adding detail to the interpretation. This is a good plan for work that is continuing, a few detail lines now, a few a little later, as may be the case in some areas, particularly on land. However, when shooting programs are large and infrequent, as in many offshore areas, the progress in seismic exploration comes into the picture. Then, especially when the time between programs has been more than a year, it may be advisable to interpret the new data, use it as a framework, and adjust the old to fit it. In other words, if some of the data has to be hammered into place, let it be the poorer data. Why force better data to fit a framework of poor data, just because the poor has already been interpreted? Well, there often are reasons—deadlines, manpower shortages, expense, other projects that must be worked on. You have to do the best you can under the conditions at the time.

And, if anybody wonders whether data really needs to be hammered into place, take it from me, it does. This seismic data isn't that absolute. It is often ambiguous, confusing, misleading. As in the saying, we're a long way from our work. We're on top of the ground, and investigating places a mile or so underground. We just send a sound down there and hope it brings reliable data back with it.

So, when there is a long time between the gathering of the old and new data, and when there is time and manpower for it, and when the new data is extensive enough, it should be interpreted first, and then used as a framework, with the old data made to fit it, or if parts won't fit, those parts not used.

However, if some special aspect of the sections is being mapped, and that aspect does not appear or is not reliable on the old data, then the old data can't be incorporated in that map. For instance, if an amplitude map is being made, and old data does not have true amplitude information preserved in it, then it cannot be added to the map. But an accompanying structure map can include both old and new data.

Saving Tapes

The seismic data stored on magnetic tape was acquired at great expense. Both for preserving the data and for enhancing it later, the tapes should be well taken care of.

Magnetic recording tape for use in computers must fit high standards. There must be close tolerances, so a bit, a 1 or 0, recorded on the tape will be read by another machine as being on the same spot, and in the same relationship to other bits.

Any distortion of the tape, any stretch or shrink, will affect the precision of location of the bits. The backing of the tape is a tough plastic which is about as resistant to distortion as any material. But, although the tape is tough and the information on it is in distinct bits, it has so much information so tightly packed that the information is vulnerable to a number of hazards.

Dust or dirt on the tape can affect the ability of a machine to read all the bits. Magnetic or electrical fields can affect the magnetization. A strong enough magnetic field would erase it. Heat and rough handling can distort it.

Before data is recorded on it, the tape is usually kept sealed in its original container. After recording, the tape needs to be protected from damage.

Data is recorded in the field where the weather may be bitter cold, baking hot, humid, stormy, or dry and dusty. There may be fungus in the air, salt spray, blowing sand.

The tape should be, and is, fairly well protected during the recording. Recording is done in air-conditioned instrument rooms or truck doghouses, as most tape recording equipment requires a moderate temperature and humidity to work well. Tapes, after recording, are immediately re-sealed in plastic canisters. When a cardboard box that tapes come in is full of recorded tapes, it is sealed up and addressed to the processing center it is destined for.

The greatest risk to the tape is between the crew and the processing center. Boxes of tapes are sent from the crew. There may be only one shipment, at the end of the operation. Or there may be several during the shooting, especially if the program is large. When they are sent in, the tapes are assigned to the tender mercies of someone—shipping company, airline, client's crew boat, a person who happens to be driving that way. At this stage they are usually in the care of someone who has other concerns, of more immediate interest than caring for tapes, and who does not know what will harm them. Recorded tapes should not be:

Left out in the sun,
Left in the cold,
Left out in the rain,
Carried on deck in salt spray,
Placed near heating pipes,
Stacked under heavy weights,
Thrown around,
Checked by magnetic detectors or X-rays,
Placed beside electric motors or dynamos,
Shipped with magnetic materials,
Delayed long in reaching their destination.

All of that is too much to keep track of, so for some of it you may have to depend on luck. But the tapes are the entire product of the expense of the seismic survey, so it's worthwhile to put forth some effort to impress people with the need for extra care.

Signs like "Magnetic Tapes—Do Not X-Ray" can be put on the boxes. Insuring for high value impresses shipping companies.

In the processing center, tapes, being the center's life blood, are well cared for.

After processing, there are a number of tapes with different versions of the data on them. There are the tapes that were recorded in the field, tapes of the same data after being demultiplexed, tapes of a number of intermediate stages in processing, stacked tapes, filtered tapes, migrated tapes, and tapes with true amplitude or other special information. Normally, not all these tapes are saved. Those that are not to be saved are wiped clean of the information that is stored magnetically on them, and re-used by the data processors. Most companies keep either the field tapes or the demuxed tapes. This original information is kept for possible reprocessing by techniques that may later be developed. Tapes with stacked data but no filtering are also usually kept. That saves the stacking—the main processing that has gone into them. New playbacks can be made from them as needed, with any filters that may be wanted. Any other processing that, like migration, starts with stacked data can later be applied to them. Some companies may also keep migrated tapes or some of the intermediate tapes.

Tapes that are to be kept are best preserved by a tape storage facility (Fig. 11-3). The storage facility may belong to a tape storage company or to the oil company. When stored, the tapes may be cleaned and tension wound. The tension winding makes the tapes uniformly tight all around the roll, eliminating one cause of distortion. The cleaning and tension winding may be done about once a year. The storage facility maintains an optimum temperature and humidity, and can index the tapes, and get requested reels out of storage for reprocessing, etc.

If for some reason it is not practical to leave the tapes in a specialized tape storage, then a simple rule of thumb is, keep them in an

Fig. 11-3 Tape storage   *Credit: Indel-Davis Inc.*

air-conditioned room. The temperature and humidity that are most pleasant to people are also good for tapes.

Why should magnetic tape be so much like people? I suspect it is because the development of magnetic tape recording took place in comfortably air-conditioned laboratories. The things that experimentation showed to work best were those that worked best in those conditions. If all the inventing along that line had been done by a guy with poor blood circulation who kept his lab fairly hot, you can bet that the tape recording equipment would be somewhat different, and that people in processing centers would complain about the hot rooms they had to work in. A visit to a tape storage is pleasant.

CHAPTER **12**

General
Considerations

There are some other factors that influence seismic exploration in addition to the straightforward seismic process, other things that a person encounters in dealing with geophysics.

Proprietary and Participation

Two variations on the shooting and ownership of seismic data have become fairly popular. The simpler is called proprietary, speculation, or non-exclusive data. A contract company shoots an area at its own expense and then sells copies of the sections, tapes, etc., to as many companies as will buy them. The survey is offered as a unit or in smaller portions (Fig. 12-1).

The participation survey does not take as much money to start, but has the disadvantage of complexity. A contractor forms a plan for a program of shooting, and presents it to the industry. Oil companies choosing to participate each pay a portion of the cost. The participating companies have a say in the plan, so it is usually modified after being subscribed. The contracting company that proposed the plan conducts the survey and gives copies of the results to the participants.

After this initial phase, the data can also be sold to other companies. There is usually a provision for some of this money to be distributed among the original subscribers or to be used to obtain additional data. At some later date, when sales are less likely, all proceeds from future sales may go to the originating contract company.

The effort involved in several companies' cooperative decisions is considerable, but those companies do receive data that better fits their needs than if they had not had a part in the decisions.

Survey Problems

Surveying—just knowing the locations of points—is a never-ending source of problems in geophysics. It is usually taken for granted, and

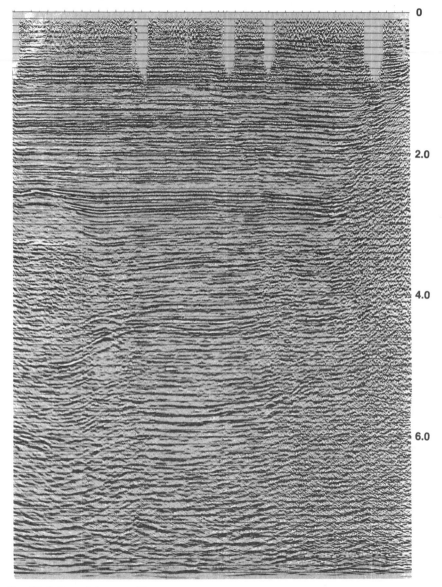

Fig. 12-1 Proprietary section
           Credit: Petty-Ray Geophysical Division, Geosource Inc.

assumed to be correct when interpretation is being done. The accuracy of a theodolite survey, or shoran, or satellite, is well known. But this is its accuracy when everything goes right. Surveying has become more accurate in recent years, but the need for precision in surveying has also increased.

On land, there are checks that can be made after a line has been surveyed. Metal tags on bushes, piles of dirt by old shot holes, cleared paths in the jungle can be found later and re-surveyed if necessary.

At sea there are no marks left after a line has been shot, so the locations of that line of data can't be re-surveyed. It can only be re-shot, that is, a new line shot at what is supposed to be the same place. There is one kind of check that can sometimes show that there is an error in 2-D shooting. Where two seismic lines cross, if a clearly recognizable reflection appears deeper on one line than on the other, something is wrong. That can't really be the intersection, or the reflections would be at the same depth. Or a prominent feature—fault, reef, salt dome, etc.—may form a coherent picture on some lines, and maybe another coherent picture on another set, but out of accord with the first. If shifting one set of data makes the picture fit together better, then probably one set is in the wrong location. Which set? The misfit just indicates the problem, but doesn't give any clues to which set should be moved. You can tell only if one set of data is firmly tied to wells, water depths, landmarks like buoys or islands, etc. And a tie of seismic data to subsurface well data must be considered suspect. The two kinds of data are too easy to rationalize into an apparent fit. Only if there is a definite mis-tie can you know. And then you only know that something is wrong.

Probably the best way to handle survey data is just—very carefully. That is, whatever survey system you use, give it a good bit of attention, don't take it for granted. If you are involved in the shooting, talk to the surveyors, get them to explain their problems. If you are in charge of the operation, spend the money for extra equipment, extra checks on the survey, etc., that they recommend. Above all, pay attention to the surveying.

Seismic Reliability

Seismic exploration is expensive, difficult, requires much time, needs the efforts of specialists to interpret it, and then maybe results in a dry hole. Why is that?

There are a few sayings to discuss in that connection. "The seismic was wrong, we hit the Devonian 50 feet low." "But we need a location over here for the lease situation." "If it won't produce here, it isn't big enough anyway."

The sayings all have to do with a philosophy of seismic exploration and what it will, or should, do, and what it won't do. Drilling wells is expensive, and you don't want to drill dry holes. Even more, you don't want to miss a field, decide it doesn't exist, and leave it for somebody else to find. To reduce these risks, the people involved in exploration decisions all need to understand each other, so oil fields aren't missed just through a communication problem.

That sounds kind of to one side of the problem, but it does indeed tie in with seismic exploration. There are several things you would like to get from a scientific, expensive method like this. Some of those things it does better than others, and some not well at all.

The one thing seismic exploration does best is to show the approximate configurations of layers of rock.

Then it is good at giving direction of dip along the line shot. That is, if a reflection on a section slants down to the east end of the section, the formation probably does dip to the east along that line. There are some situations, like rapid velocity changes, that can make even this indication wrong, but it's one of the things the seismic method generally does best.

In some areas dip direction is all that is needed. I worked an area in Montana in which I couldn't tie wells with seismic data, but could determine a point on the pay horizon that was higher than nearby points around it. Those local bumps enabled us to find oil much more consistently than we would have if I'd devoted my time to making the data tie the wells.

Amount of dip isn't nearly as well determined seismically as direction of dip. Even small differences in velocity make determining amounts of dip less reliable.

Absolute depth would be a nice quantity to obtain, but here again velocity is a big factor. Formation tops in wells might be picked wrong, but if not, if they are on the right log kicks, they can be read down to the foot. Seismic data isn't like that. First, the wiggles on a trace are long loopy things in comparison to well log curves. One seismic peak extends over about 150 feet of depth. So, in general, information can be obtained only about every 75 feet on the section, by picking a peak or a trough. That's a bit different from fine wiggles that respond to change every foot. Also, only times are read from the sections, and if they are converted to depth it must be done with velocities that are only approximate.

So we can miss in predicting depths to formations. I've often had to explain being wrong by 200 feet, and once, in the Colorado Rockies, 900 feet. In the latter case, all I could do about the error was guess that there must be an undetected fault between the two wells that I couldn't tie.

Another thing that should be pointed out is that seismic investigations can't give depths or other information about places that were not shot. This sounds simple and obvious, but locations for wells are often selected from contoured maps, without much regard for where the seismic lines are. A large feature, of course, can be seen to build up over a considerable area, so a well can be drilled anywhere on the high part, without much chance that it will turn out to be off the high. But, now that many areas are being explored for little bumps, reefs, salt collapse features, and the like, this big-feature philosophy won't work everywhere. A well in one of these areas should be drilled on a seismic line, and at the highest point on the feature. A location a quarter of a mile away could be off the feature. Then a dry hole could be drilled, presumably on the basis of the seismic data, but actually not testing it at all. Contours are just guesses where they are not on seismic lines.

Then there is the "most tolerant position" philosophy, the idea that, when several kinds of information are available, the best position to drill is one that looks good on all of them, not one that looks excellent on just one of them. Kind of modifying this to fit seismic data alone, we could say that the best location should be drilled, not one that looks marginal. The "if it doesn't produce here, it isn't big enough" idea assumes that the seismic map is either right or wrong, perfect or worthless.

Suppose there is a structure on a seismic map, large enough to contain several well locations, and that there are some non-seismic reasons to drill one of the lower ones first, rather than the highest position. But maybe the mapping is correct in that there is a structure, but doesn't have it outlined exactly right. Then the only way to be pretty sure of being on the structure at all is to drill the highest point. If that produces, development can proceed with wells in other locations. But if a low position had been drilled first and found to be dry, the whole prospect would have been abandoned. The high point on the seismic map may not turn out to be the highest in the subsurface, but it is the most likely to be somewhere on the structure. And the structure may even be as large as predicted, but not shaped just as mapped.

So seismic exploration will not do all the things that it would be nice for it to do, but some of the things it does best, pointing out directions of dip and locating locally high places, are useful to finding oil.

Be glad it isn't the other way around, a method that ties wells perfectly, but can't locate structures. Where the wells are, you already know about.

Accuracy vs. Correctness

In addition to those questions of reliability, just how accurate are seismic reflection times? Can they be given as times plus or minus, say, 10 ms, or converted to depths plus or minus maybe 50 feet?

One of the near-surface corrections of seismic sections uses an estimated velocity to correct to a datum plane. Incorrectness of the estimate can make the reflection time incorrect by a few milliseconds.

Picks of reflections are usually made on peaks or troughs. A single trace may have the peak or trough two or three milliseconds before or after an adjoining one. The rock layers don't bob up and down that much, so an interpreter smooths out the differences between a few traces to pick a time of a reflection.

If the times are converted to depths, the velocities used for the conversions are based on smoothed and estimated data.

With all this jury-rigging, if there is a good, consistent reflection following a consistent horizon, the accuracy may be something like plus or minus 50 to 100 feet in terms of absolute well ties in depth.

In another sense, relative depth, the data is considerably more accurate. That's nice, as relative depth is enough information to find oil in some areas.

But the big problem is accuracy vs. correctness. There isn't usually much point in measuring the accuracy of the minute hand on a clock if the hour hand is broken.

The correctness problems of seismic interpretation generally involve trying to stay on the same horizon. In crossing a fault or a patch of poor reflections, the interpretation may be off by how much? 500 feet? 1000 feet?

Or, in a new area with no well control, identification of horizons may be off by almost any amount, and velocities may also be poor, particularly in the deeper part of the section, where seismically determined velocity breaks down.

Sometimes a well is drilled, and by sheer chance the seismic prediction of depth to a formation turns out to be off by something like 7 feet! The danger here is that someone in management will think this is genuine precision, and expect it on later wells. When congratulated for some accident like that, I have said something like, "No, it was three hundred feet", and then, having their attention, gone into a lecture on the crudeness of seismic interpretation.

All this doesn't detract from the ability of seismic exploration to find oil. It finds most of the oil located at present. But unrealistic precision should not be expected of it.

Plus or Minus 500, or 500 Plus or Minus

There is a growing confusion about the use of the symbol "plus or minus". Perhaps it will be useful to consider the various meanings achieved by putting pluses and minuses before and after numbers.

Plus 500
Minus 500
500 plus
500 minus
500 plus or minus
500 plus or minus 25
10000 plus or minus 500
Plus 500 plus or minus 25
Minus 500 plus or minus 25
Plus or minus 500

In other words, 500 plus or minus means "about 500". But plus or minus 500 doesn't mean that. It means "either plus 500 or minus 500".

Geological-Geophysical Coordination

There has been much talk about collaboration between geologists and geophysicists. The collaboration is necessary in order to keep the two groups from drifting apart, too proud of their own specialties, unwilling to concede much to each other. Management wants coordination. Geologists want it. Geophysicists want it. But it doesn't always happen.

Yet the talk continues, and the groups continue to be more separate than efficiency calls for. The problem, of course, is that the people concerned are busy and are trying to get correct answers in their own disciplines, which is hard enough, and they just don't get around to working together sufficiently. And when a person in one specialty comes up with a plausible idea, it isn't fun to have it knocked down by somebody in another field.

A few of my experiences have been pertinent to the problem. In one office, there were a geophysical department and a geological department. The very existence of the two groups, under different bosses, made them operate as separate units. But our offices were on the same floor of the building, so we occasionally knocked heads on projects. There was plenty of formal get-together, but the most productive sessions occurred when somebody wandered into the other department carrying a cup of coffee, or when someone was excited about a prospect and wanted to tell people about it. What happened on several occasions was that we in our department had a neat picture of the area, built up after rejecting several ideas. This new idea was good, and could be defended easily. The trouble was that the geologists also had a good solid interpretation, and the two were contradictory. There would be a period when each wondered how the other could miss such obvious evidence. Then we'd argue it out, pitching out whatever couldn't be, from one viewpoint or the other, and keeping whatever held up. Eventually, a new idea would be hammered out, that might be quite different from either of the original ones, but that fit the data well. All of us came away pleased and pretty self-satisfied.

In another office, a geologist and I were assigned a long-term project. We were in adjoining offices. I'd get some new data and run into the next room. Or he'd fit it in with some wells, and run into my room. Then we were transferred to offices two thousand miles apart, and retained the joint assignment. We still worked on the project together, but by telephone and mail. We still worked well together, but something was missing.

Other times, a geologist has worked up a prospect. I've shot it, made a seismic map, and presented it back. Each of us worked separately. Then the geologist was supposed to take over and make a recommendation. This often worked well enough, but there was always the chance that one of us would bring up some simple, obvious thing that the other hadn't known about. A lot of the work would unravel. Worse, the simple things might not happen to get mentioned at all, and the interpretation would remain, based on an oversight.

With all this reminiscence, I don't have a really good way to handle the problem. If there was one, it would be in use universally by now. Much depends on personalities, the company's history, how many people are in each department, whether consultants are used, and other factors.

The one contribution I'd like to offer is that groups that work together get an esprit de corps, an us-against-the-world feeling. So the geologist and geophysicist working on a project should feel that they are in the same little group. I don't think departments should be reorganized, but maybe a geologist and a geophysicist working on the same area could be somehow identified as a unit, and discuss the work together, so they could thoroughly understand each other's view-points, even though they still might not agree. This will work best when both are well-qualified, experienced people.

This joint operation is probably better even than the assumed ultimate of having one person who is both geologist and geophysicist. The one person has more trouble being spread thin enough to work effectively on the two kinds of data, and the interaction of two minds is lost. And it is hard to keep the prejudices and way of thinking of one field from being too strong an influence in the other. There is real benefit in having two people with different outlooks work together on a problem.

Factors in Program Planning

A general discussion of program planning was given in Chapter 3, Field Operations. But there are a number of special factors in program planning that can be discussed better now, after other topics have been covered.

First, of course, is whatever the reason for the shooting may be. It might be to investigate a prospect found by the use of well data or surface geological mapping. It may be a matter of investigating scattered leases or a large permit area.

If an earlier survey is available, it will dictate some of the features of the new program. Probably, the new lines should fill in between the old ones and tie to old lines. Critical old lines shot with a poorer technique than the present ones may need to be duplicated to see if the information on them can be trusted. Large gaps in the older control may need to be filled in.

If there are wells in the area that are not tied in with seismic control, they probably should be tied. More important than just well ties, in a new area, is having some data on a producing field, to show how the kind of feature that produces there looks on a seismic section. Any time a line over production can be obtained, it can be helpful in finding oil.

Terrain and other obstacles of course need to be considered (Fig. 12-2). This is obvious, but looking at a nice white map, a person tends to forget that there may be hills, swamps (Fig. 12-3), streams, canyons, railroads, towns in the way. A topographic map helps in planning a survey. Air photos are also useful.

But whatever aids are used, even scouting the area in advance, the unexpected may crop up when the program gets underway. Permit problems and new constructions, for instance, do not appear on the

Fig. 12-2 Rough terrain

Credit: Petty-Ray Geophysical Division, Geosource Inc.

Fig. 12-3 Swamp crew

Credit: Petty-Ray Geophysical Division, Geosource Inc.

maps. Therefore it is useful for someone on the spot to have the authority to modify the program to fit conditions. To do this effectively, that person must understand the program and why it was laid out as it was, and of course should also be competent to judge requirements for seismic lines.

Usually there is no such person in the field, so the program is planned as well as it can be, and the crew shoots it if at all possible. This has led to some rather foolish lines being shot—alongside a recently cleared line or across an obstacle that was not known about or did not exist when the planning was done.

Often a line is planned in a certain way just so it can be put on paper. For instance, if lines have been shot all around an interesting-looking place, and a line through it is needed to make make the control closer, which way should the line be run? North-south, east-west, diagonally? It may not matter, but something has to be put on paper, so it is drawn, say, east-west. When the crew goes to the field, it may turn out that a new road is being built right through there in a north-south direction. The crew doesn't know there is any latitude in the assignment, so they shoot across the road work. Some geophones are set on loose, graded earth, giving poor data. Maybe a seismic cable is cut by a grader. If they had known, they could as well have shot north-south, parallel to the road work, and out of its way.

There isn't any real way around this problem, except having a competent person who understands, and maybe helped design the program, in the field. Not much latitude can be given a field crew for two reasons. Anyone down the chain of command can get afraid of the responsibility and jell the program as drawn. And, someone can misunderstand and use more latitude than was intended, maybe repeating a line that person didn't know had been shot or getting off the prospect that was to be investigated.

A program needs to be fitted to the type of feature being sought. Large structures can be hunted most economically by shooting lines along roads. Small ones, like pinnacle reefs, call for closely spaced lines through legally drillable locations. On mountain fronts, dip lines are particularly important.

Seismic programs are sometimes planned without the participation of a geophysicist. Geology, leasing, and geophysics each have requirements that must be taken into account in planning. Geological or land considerations produce the area to be investigated in the first place, and geology provides information on regional dip and other characteristics of the area. Geophysics has requirements like tied loops and multiplicity of stack, and lines long enough to achieve full stack and good migration for the entire area being investigated.

If the program is laid out, assigned to a crew, shot, and then a geophysicist is called on to interpret it, much of the geophysicist's value on the prospect will have been lost. Program planning is as geophysical an activity as interpretation.

Continuing Program

When a company has a large area to explore, for instance a production sharing contract area or a concession, there are special situations that indicate the way seismic programs should be conducted to serve their purposes best.

The fact that it is a large area, and that it is to be explored for a number of years, means that the shooting needs to be adapted to both reconnaissance and detail purposes, and that the immediate reasons for a certain shooting program should be balanced with longer-term requirements.

At the start, any data that is already available should be obtained. This will help with the regional mapping, and is more economical than new shooting.

The first shooting the company undertakes in the area may be experimental, to determine what record quality can be obtained in the area, and the seismic techniques that are best for the area. It may also determine how reconnaissance lines should be shot.

Early in the exploration, a rectangular grid pattern for seismic lines may be established. If the direction of regional dip is known, the closer lines of the grid are run along dip.

In later shooting, there may be an effort to keep to the grid pattern. This will avoid the recrossing of prospects in odd directions that so often occurs when seismic programs are allowed to "just grow". New lines parallel to lines of the original grid will give close control for less money. Crisscrossing lines are close together near the crossing point and farther apart at other places, so it is clumsy and expensive to add more data to them. A grid can be filled in with more parallel lines at any time.

Prospects once found can be evaluated by small grids especially designed for them, not necessarily parallel to the regional grid. Dip in the vicinity of the prospect may be locally in a direction different from regional dip. So a close grid is laid out with the major lines in the dip direction. In interpreting the local grid, older lines that cross it are usually ignored, as the new lines are closely spaced enough to cover the prospect thoroughly.

A step more detailed than this is to conduct a 3-D survey over the prospect. This too is laid out in the direction that is locally appropriate.

Some special conditions call for lines in special directions not parallel to grid lines. In particular, as wells are drilled, lines may be run straight from well to well. These lines should be shot with the best technique available. When a better technique is developed, they should be re-shot. They are the best testing ground, as the wells serve as a check on the seismic interpretation. Solid information between wells will also increase the information yield of the wells. These well-to-well lines can constitute a backbone for difficult interpretation situations. Once the interpretation of those lines is resolved, the other lines can be worked outward from them and tied back into them.

Development of Oil Fields

The amount of further shooting to be done on a prospect is limited by the fact that the prospect may not produce. So the shooting is limited to an amount that can establish the presence of the prospect and a fair idea of its extent.

But once a field is discovered, two things are going to happen. A lot of money will be spent, and a number of installations will be set up. There should be good technical information to guide the spending of the money. The installations can hamper or preclude further shooting in the field.

In the flush of success after discovering an oil field, people tend to feel that the seismic work has been done and was successful, and that it is time to drill delineation wells, gather production histories, and work toward the development of the field. However, in the following few years there will be a number of occasions when more information on the shape and other details of the field will be needed. Locations of development wells, platforms, and other facilities will have to be made in part by guessing, in the absence of more data. With a field partially developed, seismic data is more difficult to obtain, more expensive, and often isn't shot.

The time to do a detailed seismic survey of the field is right after it is discovered. Particularly offshore, where the rig moves somewhere else after drilling that first well, there is an excellent opportunity for detail shooting. For a detailed picture of the subsurface, as in a field being developed, people use all the technical aids that are available. As wells are drilled, a fault pattern, for instance, can be drawn based on the different levels of a horizon in different wells, and on the absence of parts of the geologic column in wells cut by faults. But the wells are only one dimensional. Guesses have to be made about what happens between wells.

At this point, a number of lines at very close spacing should be shot. The most practical thing is to shoot a 3-D survey. A three-dimensional seismic survey is the only type of investigation that gives fairly complete three-dimensional information. The direction of a fault can be determined all through the surveyed area—again if the data is good enough. This technique is the only one that is at all likely to provide a pattern of faulting that is reliable. When a new well is to be drilled, it can be positioned in the fault block desired.

Faulting is only one feature of a field that can not be established with confidence by any of the techniques available other than 3-D seismic data. Relationships of sands encountered in different wells, extents of gas accumulations indicated by bright spots, etc., can be obtained in 3-D only by a method that provides 3-D data.

If for some reason 3-D is not shot, then the lines should be at least twice as close as exploration detail lines. In any case, the shooting should be in a consistent plan to yield data about all of the field, not specific lines individually planned to answer specific questions. Tomorrow's questions won't be the same as today's. This detailed survey will then be available for future use, as when a development well shows some surprising characteristic that sets people off on a new round of investigations about the field.

The seismic expenditure on this post-discovery shooting project will be expensive, but weigh it against the many times that amount to be spent on wells, and, if the field is offshore, on platforms, production installations, etc., all of which need detailed information to guide their planning. If the seismic survey enables a platform to be set in a better location, an unneeded well to not be drilled, or a needed well to be drilled, the financial investment can appear very small.

All this doesn't mean that the seismic information will be absolutely correct. It won't. But the information in other aspects of the business isn't absolutely correct either. The additional seismic information will give the company a better chance to plan more effectively.

The seismic program will also not be the last shooting that will ever be required. New developments in seismic exploration are being made continually. If next year produces a seismic technique that yields some new kind of data, say quantity of oil or something, it may then be worthwhile to do more shooting.

Drill the Top

When a small feature is to be drilled, the well should be on the highest point possible without getting too close to the edge.

In the case of a simple anticline or reef, the highest point is probably near the center of the feature. There are several reasons for drilling the top in this case.

1. Production may be greatest there.
2. Drainage of oil from the field will probably be more complete.
3. If there are surveying errors, the central position is most likely to be on the feature.
4. If the contouring is not correct, the center is the safest place.
5. There may be surprises at the top of the feature.

Number 5 includes extra porosity by winnowing over a structure, development of a reef on the high point of a structure, trapping of hydrocarbons in other layers above a feature. If the first well is not drilled on the top, then a subsequent well should be drilled there, notwithstanding careful attic oil calculations by engineers, based on porosities and other conditions in wells on the flanks (Fig. 12-4).

Most Important Factor

The most important thing in a seismic search for oil in a specific area is—to determine what the most important factor is, and direct the effort primarily toward it.

Sounds obvious, I know. But it isn't done all that much. People tend to have pet things they consider vital, and devote their efforts to, like doing everything possible to get the very best record quality. But maybe record quality isn't the main factor in the area.

In some areas, the quality of the data is the main thing. It's hard to get good reflections, and attention must be paid to everything that can improve them. Maybe the number of lines should be reduced, so more money and effort can be spent on getting a few good lines.

In other areas, good sharp reflections come easy. An improvement would be nice, but wouldn't find more oil. Maybe money can be spent on more lines, with less processing.

In areas of small features, the most vital thing may be close spacing of lines, carefully positioned to cross locations that can be drilled.

In areas of complex folding, direction of lines—directly down dip—and best possible migration of selected sections, may make the difference between understanding the geology and totally failing to understand it.

...., THEN, WHEN YOU CROSS MY FENCE DOWN BY THE CREEK, YOU'LL BE 'BOUT ON TOP OF THE STRUCTURE. YOUR RECORDS WILL START GOIN' BAD, LOTS OF HIGH FREQUENCY STUFF. MOST OF THE OTHER FELLOWS GOIN' THRU THERE USE A WA-30 FILTER. THEN WHEN YOU 'BOUT REACH THE HIGHWAY ETC, ETC. - - -

Fig. 12-4   Credit: J.C. Knight

In highly faulted areas, two things may be necessary: best possible record quality to define the faults, and some sort of three dimensional control to determine the directions of the faults—either a 3-D survey or something less, at the very least, pairs of closely spaced lines.

In areas of many obstacles to shooting—rugged terrain, islands, platforms—the main factor may be designing a program to get the needed control by going around or shooting under the obstacles.

In areas of difficult access, logistics may be most important—getting supplies in, having spare parts available, etc.

Once, working for a company that seemed to be putting its main stress on number of profiles (shot points), a friend of mine commented that he thought seismic success was measured in barrels, not profiles.

How to Find Oil

This heading turned you on? Thought it might. And I do indeed intend to tell you how to find oil by seismic exploration.
1. First step. Have an area with good chances for oil.
2. Have the area worked on by a geophysicist in all its stages, from planning through deciding on well locations. The geophysicist needs to work on it long enough to learn its quirks, say a year or two as a minimum. The geophysicist should be competent, imaginative, optimistic, and if not directed too much, should produce some good surprises.
3. Provide plenty of opportunity for interaction between geophysical and geological personnel.
4. Make sure that the most important factor is determined, and other important factors. This isn't necessarily an early step. People just find it out as soon as they can. It may only take some conversations with people who know the ropes in the area. Or, much more likely, it may take several years of work.
5. If there are already-discovered oil fields in the area, get some seismic control—old or new—over them.
6. Shoot a program or programs that fit the area. Don't just lay them out. Get seismic expertise in the act.
7. Don't "science it to death". If you get some prospects to drill, think things out thoroughly, raise objections, but don't go through lengthy re-investigations too long. The prospect can't be proved until it's drilled, so don't expect it to be.

No guarantees, but this is the best way I know to give yourself a fighting chance to find a lot of oil by seismic methods.

Appendix

Other Sources of Information

This section takes the place of a list-type bibliography. It seems preferable to have a discussion of the references, describing what can be found in them. The references are listed more or less in order of usefulness to the person who does not want to, or is not ready to, go deeply into the technical side of geophysics, or the person who knows the technical side, and wants to learn the workaday techniques.

ENCYCLOPEDIC DICTIONARY OF EXPLORATION GEOPHYSICS, Second Edition
(Compiled by R.E. Sheriff, 1984, The Society of Exploration Geophysicists, Tulsa)

This book is essential to anyone dealing with geophysics. No matter what your activity, there will be terms that are unfamiliar to you in geophysics or related fields. It explains clearly. Its main use is as a reference, to answer "What does this word mean?" But it is also worthwhile to thumb through it. You are almost certain to find a chart or definition that you will profit from.

DICTIONARY OF GEOLOGICAL TERMS, Third Edition
(American Geological Institute, 1984, Doubleday & Company, Inc., New York)

This dictionary, while primarily geological, defines some geophysical terms. It does so from a somewhat different viewpoint from the one used in this book or the one used in the geophysical dictionary above. So it is worthwhile to have available also.

COMPANY HANDOUTS

This is the best, and in some respects may be the only, source of up-to-date information on geophysics. The handouts are mostly from geophysical contractors and suppliers. They are in the forms of advertising brochures, explanatory folders, copies of technical papers. Some of the papers are mathematical, but the other things tell about whatever the company is proud of, and tell it understandably. And they can be on subjects not normally published in magazines or books—a new type of seismic section, new cable, vehicle, etc. These handouts are most readily available at geophysical and related conventions, where many companies have booths in the same building. The handouts can also be obtained from company offices, but there is no good way to know what to ask for. It's easier to look around at a convention.

ARTICLES IN GENERAL OIL MAGAZINES

The oil magazines that are not primarily devoted to geophysics—*The Oil and Gas Journal, World Oil, Offshore,* etc. quite often have articles on geophysical topics. These articles tend to be practical and understandable to non-geophysicists. Many of the articles are well worth looking out for, and either keeping in a file or listing in an index so they can be referred to when needed.

BOOKS

A number of books on geophysical exploration have been published through the years. Some are texts useful for going into the theory of geophysics. Some cover the entire field with considerable mathematics. Some are on a more popular level.

AAPG BULLETIN
(American Association of Petroleum Geologists, Tulsa, monthly)

The *AAPG Bulletin* is, of course, devoted to articles by and of interest to geologists. However, its articles on specific areas are often illustrated with seismic sections, and may tell of seismic problems encountered in exploring the areas. These articles are quite useful if you are planning to shoot one of those areas, or a similar area somewhere else. These articles too could advantageously be filed or indexed for reference.

AAPG EXPLORER
(American Association of Petroleum Geologists, Tulsa, monthly)

This is a less technical publication for geologists. It contains articles of a popular nature that would not be likely to be in the more technical *AAPG Bulletin*. It often includes articles about geophysical operations.

GEOPHYSICS
(Society of Exploration Geophysicists, Tulsa, monthly)

Geophysics is a magazine intended for geophysicists. However, as things have worked out, it is primarily for, and by, research geophysicists, so it is mostly pretty mathematical.

The advertising in *Geophysics* is something of a short version of the company handouts, and is useful in the same way they are.

THE LEADING EDGE (full title: *Geophysics: the Leading Edge of Exploration*)
(Society of Exploration Geophysicists, Tulsa, monthly)

This is the SEG's popular-type magazine. It has some articles about seismic methods, but without the mathematics that appears in *Geophysics*.

GEOPHYSICAL PROSPECTING
(European Association of Exploration Geophysicists, The Hague, Netherlands, published eight times a year)

This magazine is very like *Geophysics*.

FIRST BREAK
(Blackwell Scientific Publications, Oxford, England, with the EAEG, monthly)

This is the EAEG's popular-style magazine.

LOCAL GEOPHYSICAL AND EXPLORATION SOCIETIES

These societies publish many useful articles on the exploration of their areas. The articles are useful to people in other areas that have similar exploration situations. The publication is in the form of books (often loose-leaf) that are guidebooks to field trips, collections of papers from technical meetings, or other group projects. Some of the societies also publish magazines.

Educational Requirements

There is a confusion in the minds of some management and personnel people on the education required for geophysical work. The confusion is caused in part by the many aspects of geophysics with their many different requirements, and in part by the general hiring problem faced by personnel departments.

The personnel hiring problem is the same as the one in hiring secretaries. If you ask the personnel department to find a secretary for you, they will ask, "How many words per minute?" For many secretarial positions, that question is almost useless. You want a secretary to keep the files straight, find things when you need them, remind you of appointments, type a few letters, and occasionally type a long report. The only time typing speed makes any difference is in the report, and even then it isn't nearly as important as correct typing, good proofreading, knowledge of spelling, and good arrangement of text on the paper. The ability to proofread is surely the most useful quality in getting a report out. So why do personnel people ask about typing speed? Because they can measure it. It's a thing people can be graded on, and it helps excuse the failure to judge other factors. The failure isn't their fault, as the other factors are hard to measure.

In seismic occupations, the comparable question is, "What degree do you need?" Again something that is easy to determine. Without thinking much, you may answer, "A Master's in geophysics, unless we can get a Ph.D." This not only isn't the best answer, it may be seriously wrong.

The different aspects of geophysics with, more or less, their different requirements are:

FIELD WORK. The main ability needed for this is to be able to make equipment work somehow when things break down. A field person needs to be able to improvise. Experience in keeping equipment working since childhood helps. The usual way to do that is to grow up on a farm. Maybe an adolescence spent fooling with old cars would be a second best. If you can get people like this, you are way ahead. This applies while they are in helper positions in the field, and also later, when they become field supervisors, heads of departments, presidents of companies.

RECORDING. Recording equipment is electronic. For operating it, a radio ham or computer hobbyist might be a good type, and/or a person with an electronics degree.

PROCESSING. Data processing is done with computers. Education in all aspects of computers and in the applicable mathematics is useful. Experience with personal computers helps.

INTERPRETING. Here, for the first time in this list, is a field that calls for education directly in geophysics. A variety of geophysics courses is needed. A degree in geophysics indicates that the courses have been taken. Courses in geology are necessary. They should include structural geology, stratigraphy, and historical geology.

In all these parts of geophysical exploration, I have listed the background that should be looked for when hiring a person. Note that degree level was not mentioned. If a company selects for people of high degree level, it is selecting in part for people who like to go to school, and not selecting for practical tendencies. The degree should be considered secondary to other factors.

Now we'll turn the thing around. Instead of a company's viewpoint in hiring, consider an individual's position. When a person is thinking about whether to get more education, the answer should almost always be to do so. Certainly, a person in any of the parts of the geophysical business will profit from a course in geophysics. The same goes for a course in geology, petroleum engineering, electronics, computer programming, etc. You will do your work better if you know your specialty better, and if you know related fields outside of yours. And, while working in a rapidly-changing field, it is essential to put forth efforts to keep up with new developments. Summing up, it is great for a person to get more education, but it is hazardous for a company to hire people on the basis of formal education, particularly based on degree level.

Yet another aspect of geophysical education is that gained by experience. Anyone planning to work in one of the specialized fields of geophysics would benefit greatly by acquiring some experience in that field, and also in other fields, while still going to school. Summer vacations from college spent working on a seismic crew or in a processing center not only build up some experience, but also give the courses in college more meaning. A person who plans to make a living in data acquisition, processing, or interpreting should, if possible, spend some time on a crew—better yet, on both a land crew and an offshore crew. The person also needs experience in a processing center and in an interpreting office. Any of this summer work will necessarily be low-level, just following orders. Fine, that is ideal for gathering experience.

Then, after being hired on a regular basis, there is still a need to gain a variety of types of experience. Some companies put new geophysical employees through a training period that includes duties on various types of operation. With a company that does not offer such a program, or in later years after completing that program, the best way to get more variety of experience is to be willing, and known to be willing, to fill in wherever there may be a need.

Still another educational need extends all through a geophysical career. It is the need to keep up to date in a rapidly-developing field. In present-day geophysics there is no such thing as learning the field in school and therefore being well enough educated in it for many years' work. The best sources of information on the new techniques will depend on the person's specialty. For a research geophysicist, and maybe for a processor, articles in Geophysics magazine may be best. For field work, the contractors' brochures and discussions with other geophysicists are probably best. An interpreter who becomes involved with crew selection and supervision may need all these methods.

We hear about the sad state of people who must be retrained for another job. In a technical field, a person needs to be in a constant process of retraining. That's a lot of study over a lot of years, but a type of work that didn't change much wouldn't be as exciting. Geophysical exploration involves adventure.

What Is a Geophysicist?

I was sitting at my desk minding my own business, doing whatever a geophysicist does for a living, when I heard voices in the hall. Apparently a secretary or draftsman was showing a new employee around the company offices. She must have just arrived at our floor and pointed to our office.

"That's the geophysical department."

Pause, while she apparently decided a little more explanation was in order.

"They're geologists."

Another pause. Not quite right yet.

"Special kind of geologist."

Pause.

"The real geologists are upstairs."

For Non-Oil People

This book was aimed at a certain audience, one that has reason to learn something—about as much as one book will hold—about seismic exploration. But other people, who don't have this incentive, may ask you what it's all about. The following segment is for you to show them. It's the explanation that might be given in the process of asking permission to shoot on a farmer's land, a farmer who was interested in what was going on, beyond the usual "We want to make some tests". Or to a relative who, to a geophysicist's delight and amazement, shows an interest in what a geophysicist does for a living.

So, on the following pages is a brief explanation for people who are interested, but shouldn't be exposed to more than their interest calls for.

And, if you are a geologist, production man, or other specialist, pardon me for getting into your field. But this is for people who don't know about the oil business.

How We Find Oil by Seismic Exploration

Oil, petroleum, crude oil, whichever name you like, is found underground, often deep underground. But it doesn't occur in a stream, pool, or lake of oil.

Oil occurs in rocks, that is, in any open spaces there are in the rocks—between the sand grains of a sandstone, in the cracks in a shattered rock, in the little cavities in a limestone. The grains of a sandstone are packed tightly together. Some buildings are made of sandstone, and they don't seem to have any open spaces in the blocks. But the sand grains are mostly round, so there are tiny spaces between them. A sandstone building may

Fig. 1 Little chunk of rock, with water or oil (black) between grains

sop up rain, and stay wet in spots for days as the water seeps out of it.

Much of the earth is covered with sedimentary basins, places that once were seas. Through the years dirt, sand, etc. were washed from the land into the seas, forming layers of sediment, whatever settled down through the water. As more layers were added, their weight compressed the earlier layers underneath, turning them from mud, sand, etc. into rock. The gases and liquids were squeezed out of the rock, except in the remaining pore spaces left by the shapes of the hard grains. The fluid left in the pores was mostly salt water from the sea.

Fig. 2 Old sea in cross section, with sediment under the water

The sedimentary basins have, since they were formed, been warped, deformed, distorted by various forces in the earth—volcanoes, faulting (which we feel as earthquakes), salt flow, continental drift, mountain uplift, etc. The basins may now be dry land, or they may still, or again, be under the sea.

Fig. 3 No longer sea—beds distorted

With the rocks wet with salt water, there may have been impurities in the salt water like bits of plant and animal matter. With time, pressure, and heat organic matter became petroleum and natural gas, scattered through the wet rock. But oil floats on water, so the oil, through the ages, crept upward to the top of the layer of wet rock it was in. If the rock layer was later tilted, the oil would work upward clear to the top of the ground and become an oil seep, if its path was uninterrupted.

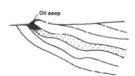

Fig. 4 Detail of Fig. 3

As a result of the deformations, though, there may be some local high spots, where the layer was pushed up. If oil floats into one of them, it can't get out. It is trapped. This is an oil trap. Gas is lighter than oil, so any natural gas in the rock will float to the top of the oil.

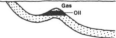

Fig. 5 Oil trap

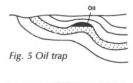

Fig. 6 Oil and gas trap

Seismic exploration is a means of looking for oil and gas traps. If you shout, the sound may echo back from a cliffside or building in the vicinity. If you count seconds to see how long a time it took for the sound to return to you, you can figure

Fig. 7 Echo

how far away the cliff is. Sound travels about 1000 feet a second in air, so multiply 1000 by the number of seconds you counted. That shows how far the sound traveled. If it took four seconds, the sound went 4000 feet in going to the cliff and back. So the distance just to the cliff is half that far, 2000 feet.

That's what we do with seismograph, but with more careful measurements. We make a sound and use very sensitive instruments to detect echoes. We aren't interested in learning how far away a cliff or building is, but do care how far it is to a layer of rock deep down under our feet. We try to make most of the sound at a pitch that penetrates rock well, which happens to be very low tones, some of them lower than people can hear.

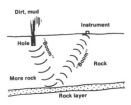

Fig. 8 Underground echo

So all we do is go someplace, usually a place selected by a geologist who has worked with information from wells. Then we produce the sound, have instruments detect it and record it for later reference, go someplace else and do the same, etc. Finding how deep some rock layer is at a number of places will show where it is pushed up. This pushed-up place may be an oil trap, so it may be a good place to drill for oil. We can't tell for sure, but the odds are lots better than drilling just anywhere.

Briefly, in practice —

A plan of places to be investigated is made up,

Sounds are made a little way underground or on the ground or in the water,

The echoes of those sounds are recorded,

The recordings are made into a cross section of the earth to a depth of a mile or so,

These cross sections are used to predict where oil is likely to be found.

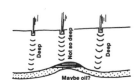

Fig. 9 Three underground echoes—not at same time

We do seismic work on land and at sea. Sedimentary basins can be either place.

On land, the work goes like this, but with variations to fit local conditions.

A seismograph crew is assigned a certain area to "shoot" (acquire seismic information in).

A permit agent visits the owners of the land, and asks permission to shoot.

A surveyor with an instrument on a tripod sights through the instrument to a painted board held upright by an assistant. They do this here and there to survey the locations to be shot, so the locations are known and can be put on a map.

Fig. 10 Surveying

People lay out a long cable, plant little geophones (fist-size or smaller) on the ground to detect the sound, and connect them to the cable.

A truck with a super tape recorder in it is connected to the cable, and through it, to the geophones.

Some way of making the sound is employed. A truck with a heavy weight under it may press the weight against the ground and vibrate it. A drill on a truck may drill a hole in the ground in which an explosive can be fired. A truck may drop a weight a few feet.

Fig. 11 Drilling

The sound and its echoes are recorded for about six seconds on magnetic tape, the way music and speech are recorded by a home tape recorder.

The "shot points" where the sounds are made are in long straight lines. There is usually considerable running back and forth to keep cable and geophones advancing to new shot points.

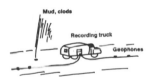

Fig. 12 Shooting and recording

Permit agent, surveyor, recorders, usually arrive on different days. The lines to be shot are planned in advance and the various parts of the crew try to arrange their work so no part has to wait on another.

At sea a seismic operation is quite different. The crew is all together on one big boat, 100 to 200 feet long. Seismic boats are seaworthy enough to cross oceans, or to stay at sea, shooting, for a month or more without needing additional supplies.

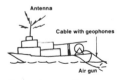

Fig. 13 Seismic boat

The surveying is done, not by sighting through an instrument, but by radio. The radio surveying may be accomplished by measuring the times in which radio signals travel from two special broadcasting stations on land to the boat. This gives the distance from each station, so the location of the boat can be determined.

Or the radio surveying may use earth satellites, with a receiver on the boat picking up the satellite's broadcast beeps. From the precisely-known orbit of the satellite, a computer on the boat calculates the boat's location. In either case, one or two people sit in a cabin with radio equipment, keeping it working.

Instead of people laying out cable and geophones, the boat tows a cable more than a mile long, with the geophones built into it. There is a large reel on the boat, to wind the cable on when it is not in use.

For shooting, the boat usually tows a number of air guns. The air guns are chambers that hold compressed air from compressors on deck, and release it suddenly in the water. The pops of air are the sound that is then reflected from rock layers far below the sea bottom, and received by geophones in the cable.

In a large cabin is the big tape recorder that records the information from the geophones. There is lots more room on a boat than on a truck. The reflected pops are recorded on magnetic tape.

There are skipper, boat crew, cooks, bunks, etc. on the boat. Seismic field work on land isn't practical at night, but offshore it goes on 24 hours a day. The boat rolls and pitches 24 hours a day, too.

The magnetic tape, recorded either on land or offshore, is taken to a computer center. Sophisticated processing is done in a number of steps to improve the quality of the recordings. Recordings of many shots are put together to make a seismic

Fig. 14 Reel of magnetic tape

section, a sort of pictorial cross section of the earth. The section may be on paper or on a TV-type screen.

An interpreter sits down with the sections—one for each straight line that was shot. The interpreter picks out the echoes from a rock layer and determines how deep it is at a number of points. This information is put on a map, with contour lines drawn to show the high and low places for the rock layer. If the contours show a good high place that might be an oil trap, the interpreter may recommend that the oil company drill a well there.

An oil derrick is set up to drill a well on land, often a derrick that is built on a truck-trailer and, at the location, raised to an upright position. When a well is to be drilled offshore, the rig is floated to the location, where it is secured in place by anchors, or raised to stand on long steel legs.

After all this, more often than not, the well doesn't find oil or gas. So the would-be oil well is plugged and abandoned. It is called a dry hole or even a duster, although it isn't dry or dusty. There just isn't oil or gas in it. But remember that salt water? If there was a market for salt water from ancient seas, the drillers of "dry" holes would be happy.

And while we're at it, in spite of the insistence of television and movies, a well that discovers oil isn't a "gusher" these days unless someone has been inept. Big valves called blowout preventers can be closed to hold the oil back. In the old days, gushers wasted much oil and were fire hazards. The most spectacular sight nowadays is in testing a well, when oil is allowed, under control, to squirt sideways over a mud pit. Any gas is burned off as a safety measure. So a good well may for a few minutes have a dramatic 40-foot flame shooting out from it.

If an offshore well discovers oil, the rig is

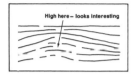

Fig. 15 Seismic section

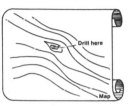

Fig. 16 Contoured map

moved away, a permanent platform is built on
steel piling, and equipment to handle the oil is
placed on the platform.

By the way, an offshore platform is the place
for deep-sea fishing. Barnacles and things grow
on the piling, little fish come to eat them and hide
behind the piling, bigger fish come to eat them.
The fishing in the area improves dramatically.

Fig. 17 Fishing

Both on land and at sea, if a well discovers oil,
then the oil field is developed. Other wells are
drilled around it to the limits of the field, that is,
until dry holes are drilled. We don't know how big
the field is until the dry holes around it are
drilled.

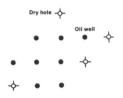

Fig. 18 Developed oil field

Index

Acceleration-cancelling hydrophone, 86
Accordion: type of display, 248
Accuracy: compared with correctness, 311–312
Acoustic surveying: for marine area, 84
AGC: automatic gain control, 115
Air compressor: for air gun, 90
Air drill: for shot hole, 60
Air gun: land, 69; marine energy source, 89–91; in tuned array, 90–91; for velocity survey, 198; on LVL crew, 290
Air-hammer drill: for shot hole, 60
Alias: apparent different frequency, 111; anti-alias filter, 138
Alidade: surveying instrument, 55
Ammonium nitrate: charge in shot hole, 65
Amplitude: of reflected wavelet, 48; determined by interface, 146; in Fourier analysis, 152; on time slice, 165–166; displayed in color, 168–169, 274; true amplitude section, 192; in synthetic sonic log, 203–204; anomaly, 274–278; effect of velocity on, 276
Analog: representation of data, 109–111; anti-alias filter, 138; early tapes, 138, 299
Anchor: to hold charge down shot hole, 63
Angle: of incidence and reflection, 6–7; of fault, 267; of reflection in seismic lithology, 280; see also Critical angle
Antenna: for survey base station, 83

Anticline: effect of migration on, 173–175; shown by contours, 228; sheepherder, 254
Array: of air guns, 90–91
Assignment map: for field crew, 54
Attribute: seismic, 252, 274
Auger drill: for shot hole, 60
Autocorrelation: trace correlated with itself, 144; to determine wavelet, 151
Automatic gain control, 115
Automatic volume control, 115
AVC: automatic volume control, 115

Balancing: of cable, 86–87
Bandpass filter: to eliminate frequencies, 138
Base line: between survey base stations, 82; crossing of, 82
Base map, 227
Base station: in surveying, 81–83; base line between, 82; antenna for, 83
Benchmark: to survey from, 57; placed by surveyor, 58
Bentonite: to plug shot hole, 63
Bid: for field work, 54
Bin: for CDP data, 97
Binary: number system, 111–113
Binary-coded decimal: number system, 113
Binary gain: recording system, 115; for synthetic sonic log, 204
Bit: on drill stem, 59–60; representing data on tape, 116; checked by parity, 117
Blasting cap: for charge in hole, 63